# Data, Engineering and Applications

Rajesh Kumar Shukla · Jitendra Agrawal ·
Sanjeev Sharma · Geetam Singh Tomer
Editors

# Data, Engineering and Applications

Volume 2

 Springer

*Editors*
Rajesh Kumar Shukla
Department of Computer Science
and Engineering
Sagar Institute of Research & Technology
(SIRT)
Bhopal, Madhya Pradesh, India

Sanjeev Sharma
School of Information Technology
Rajiv Gandhi Technological University
Bhopal, Madhya Pradesh, India

Jitendra Agrawal
School of Information Technology
Rajiv Gandhi Technical University
Bhopal, Madhya Pradesh, India

Geetam Singh Tomer
THDC Institute of Hydropower
Engineering and Technology
Tehri, Uttarakhand, India

ISBN 978-981-13-6353-5          ISBN 978-981-13-6351-1   (eBook)
https://doi.org/10.1007/978-981-13-6351-1

Library of Congress Control Number: 2019931523

This Springer imprint is published by the registered company Springer Nature Singapore Pte Ltd.
The registered company address is: 152 Beach Road, #21-01/04 Gateway East, Singapore 189721,
Singapore

# Contents

# About the Editors

**Dr. Rajesh Kumar Shukla** is a Professor and Head of the Department of Computer Science and Engineering, SIRT, Bhopal, India. With more than 20 years of teaching and research experience he has authored 8 books and has published/presented more than 40 papers in international journals and conferences. Dr. Shukla received an ISTE U.P. Government National Award in 2015 and other various prestigious awards including some from the Computer Society of India. His research interests include recommendation systems and machine learning. He is fellow of IETE, a senior member of IEEE, a life member of ISTE, ISCA, and a member of ACM and IE(I).

**Dr. Jitendra Agrawal** is a faculty member in the Department of Computer Science & Engineering, Rajiv Gandhi Proudyogiki Vishwavidyalaya, Bhopal, India. His research interests include data mining and computational intelligence. He has authored 2 books and published more than 60 papers in international journals and conferences. Dr. Agrawal is a senior member of IEEE, life member of CSI, ISTE and member of IAENG. He has served as a part of the program committees for several international conferences organised in countries such as the USA, India, New Zealand, Korea, Indonesia and Thailand.

**Dr. Sanjeev Sharma** is a Professor and Head of the School of Information Technology, Rajiv Gandhi Proudyogiki Vishwavidyalaya, Bhopal, MP, India. He has over 29 years of teaching and research experience and received the World Education Congress Best Teacher Award in Information Technology. His research interests include mobile computing, ad-hoc networks, image processing and information security. He has edited proceedings of several national and international conferences and published more than 150 research papers in reputed journals. He is a member of IEEE, CSI, ISTE and IAENG.

**Dr. Geetam Singh Tomer** is the Director of THDC Institute of Hydropower Engineering and Technology (Government of Uttarakhand), Tehri, India. He received the International Plato award for Educational Achievements in 2009. He completed his doctorate in Electronics Engineering from RGPV Bhopal and post-doctorate from the University of Kent, UK.

Dr. Tomer has more than 30 years of teaching and research experience and has published over 200 research papers in reputed journals, as well as 11 books and 7 book chapters. He is a senior member of IEEE, ACM and IACSIT, a fellow of IETE and IE(I), and a member of CSI and ISTE. He has also edited the proceedings of more than 20 IEEE conferences and has been the general chair of over 30 Conferences.

# Part I
# Big Data and Cloud Computing

# Efficient MapReduce Framework Using Summation

Sahiba Suryawanshi and Praveen Kaushik

## 1 Introduction

BigData can be defined as huge quantity of data, in which data is beyond the normal database software system tool to capture, analyze, and manage. The data is within the limits of three dimensions which are data volume, data variety, and data velocity [1]. Primary analysis data contains surveys, observations, and experiences; and secondary analysis data contains client information, business information reports, competitive and marketplace information, business information, and location information that contains mobile device information. Geospatial information and image information contains a video and satellite image and provides chain information containing rating and vendor catalogs, to store and process this information that is done by BigData. To process this variety of data, the velocity is incredibly necessary.

The major challenge is not to store the big datasets in our systems, but, to retrieve and analyze the large data within the organizations, that too, for information stored in various machines at completely different locations [1]. Hadoop comes in a picture in these situations. Hadoop has been adopted by many people leading companies, for example, Yahoo!, Google, and Facebook along with various BigData programs, for example, machine learning, bioinformatics, and cybersecurity. Hadoop has the power to analyze the info very quickly and effectively. Hadoop works best on semi-structured and unstructured data. Hadoop has MR and Hadoop distributed file system [2]. The HDFS can provide a storage for clusters, and once the info is stored within the HDFS then it breaks into number of small pieces and distributes those small items into number of servers that are present within the clusters, wherever each server stores

S. Suryawanshi (✉) · P. Kaushik
Department of Computer Science and Engineering, Maulana Azad National Institute of Technology (MANIT), Bhopal, India
e-mail: sahiba686@gmail.com

P. Kaushik
e-mail: kaushikp@manit.ac.in

© Springer Nature Singapore Pte Ltd. 2019
R. K. Shukla et al. (eds.), *Data, Engineering and Applications*,
https://doi.org/10.1007/978-981-13-6351-1_1

these small pieces of whole information set and then for each piece of information a copy stored on more than one server, this copied information set will be retrieved once the MR is process and within which one or a lot of *Mapper* or *Reducer* fails to process [3].

MR appeared as the preferred computing framework for large processing because of its uncomplicated programming model and the execution is done in parallel automatically. MR has two computational phases, particularly *mapping* and *reducing*, that is successively carried through many *maps* and *reduce* tasks unlike. The *map* reads input data and manages to create *<key, value>* pairs depending on the input data. This *<key, value>* pairs are the intermediary outputs within the native machine. Within the *map* phase, the tasks begin in parallel which generates *<key, value>* pairs of intermediate data by the input splits. The *<key, value>* pairs are kept on the native (local) machine and well ordered into various data partitions as one for each *reducer* phase. *Reducer* is liable for processing the intermediary outcomes which receive from various *mappers* and generating ultimate outputs within the *reducer* phase. Every *reducer* takes their part of information from data partitioning coming from all the *mapper* phases to get the ultimate outcome. In between the *mapper* phase and the *reducer* phase, there is a phase, i.e., *shuffle phase* [4]. During this, the info created at the *mapper* phase is ordered, divided, and shifted to the suitable machine execute the *reducer* phases. The MR is performed over a distributed system composed of the master and a group of workers. The input is split into chunks that are allocated to the *mapper* phases [5]. The *map* tasks are scheduled by a master to the workers, which consider the data locality. The *map* tasks provide an output which will be divided into a number of pieces according to the number of *reducers* for the job. Record with the similar intermediary *key* should go to the same partition so that it will guarantee the correctness for the execution. All the intermediary *<key, value>* pairs' partitions are arranged and delivered with the task of *reducer* that needs to be executed. By default, the constraint of data locality does not take into consideration while doing the scheduling tasks for the *reducer*. As the result, the quality of data at the *shuffle* phase, which needs to transfer via a network, is significant. Using tradition, a hash-based function which is used for partitioning the intermediary data in *reducer* tasks is not traffic-efficient because topologies of network and size of data corresponding to each key are not considering it. Thus, this paper proposes a scheme which sums up the intermediate data. The proposed scheme will reduce the data size that has to be sent to the *reducer*. By reducing the size of data, the network traffic at reduce function will minimize. Even though the combiner also performs the same function, the combiner operates on the generated data by *map* task individually which thus fails to operate between the multiple tasks [4]. For summing up the data, it put summation function. Summation function can be put in both either within the single machine or among different machines. For finding the best suitable place for summation function, it uses distributed election algorithm. At the *shuffle* phase, the summation function will work simultaneously, and then it removes the user-irrelevant data. In this, the data of volume and traffic is reduced up to 55%, and then it sends to the *Reducer*. It is more efficient way to process the data, for those jobs which have hundreds and thousands of key ends, and each of the keys is associated with number of values.

The rest of the paper is organized as follows. In Sect. 2, we review recent related work. Section 3 provides the proposed model. Section 4 analyzes the result. Finally, Sect. 5 concludes the paper.

## 2 Literature Survey

In this section, different techniques for optimization generally applied in the *MapReduce* framework and BigData are discussed. The paper also discussed the attributes of various techniques for optimization and how BigData processing is improved by these techniques.

In [6], the author examined that whether the network optimizing can make a better performance of the system and realize that utilizing the high network and low congestion in a network; good performance can also be achieved parallel with a job in the optimizing network system. In [7], the author gives purlieus, a system which allocates the resources in MR, which will increase the MR Job's performance in the cloud, via positioning intermediary data to the native machines or nearer to the physical machines. This reduces the traffic of data within the *shuffle* part produced in the cloud data center. In [8], paper designs a good key partition approach, in which the distribution of all frequencies of intermediate keys is watched, and it will guarantee that the fair distribution in *reduce* tasks, in which it inspects the partitioning of intermediary key and distribution of data with the key and its respective value among all the machines of *map* to *reducer*, for the data correctness, is also examined. In [9], the author gives two effective approaches (load balancing) to skew the data handling for MR-based entity resolution. In [10], the author proposes MRCGSO mode; it adjusts very well with enlarging data set sizes and the optimization for speedup is very close. In [11], the author relates the parallel and distributed optimization algorithm established on alternating direction method of a multiplier for settling optimization problem in adapting communication network. It has instigated the authorized framework of the extensive optimization problem and explains the normal type of ADMM and centers on various direct additions and worldly modifications of ADMM to tackle the optimization problem. In [12], the author pays attention to accuracy using cross-validation; the paper gives sequential optimize parameters for plotting a conspiracy of accuracy. In [13], the author gives a method to initiate Locality-Sensitive Bloom Filter (LSBF) technique in BigData and also discusses how LSBF can be used for query optimization in BigData. In [14], the author initiates the optimization algorithms using these rules and models can deliver a moderate increase to the highest productivity on inter-cloud and intra-cloud transfers. Application-level transfer adapts parameters just like analogousness pipelining, and concurrency is very needful mechanisms for happening across data transfer bottlenecks for scientific cloud applications, although their optimal values juncture on the environment on which basis the transfers are generally done. By using actual models and algorithms, these can spontaneously be optimized to achieve maximum transfer rate. In [15], the author offers an algorithm for cache consistency bandwidth optimization. In this perspective, the user data shift is optimized without consideration of the user expectation rate. The debated algorithm differentiates with trending data transmission techniques. In [16], the author recommends improved computing operators focused on smart grids optimization for BigData through online. This settles a problem of generic-constrained optimization by utilizing a module based on the MR framework, which is a trending computing platform for BigData management commenced at Google. Online answer to urged optimization problems is a necessary requirement for a secure, reliable smart grid operation [17]. Many authors proposed methods for optimizing MR, but very less work is done for optimizing MR by reducing the traffic generated, while the data is sent to *reducer* phase. So we proposed a method which is based on distributed summation.

## 3    Proposed Method

In Hadoop MR, normally the *reduce* task resides in different machines/racks. As the massive intermediate data goes to *reduce* task, it will create heavy traffic in the network. By analyzing the intermediate output, we saw that there are redundant *<key, value>* pairs. So we can sum up all the similar *<key, value>* pairs before sending it to the *reducer*, which will minimize the quantity of intermediate data. The *summation* can be done in both either within the machine or among different machines.

Summation at single machine: The *summation* function can put at each machine in which it will sum up the similar *<key, value>* pairs within the same machine before it is sent to the *reducer*. As a result, the data from each machine will minimize by *summation* function that will minimize the traffic too.

Summation among different machines: When the *summation* function put at each machine, it will reduce the size of data. But for massive data, it needs many *mappers* and *reducers*. There may chances that at *reducer* there might be number of inputs even though the inputs are already minimized by *summation*. But due to number of *map* functions, it will also create traffic at *reducer*. For that, it will sum up data among different machines.

In Fig. 1a, it needs to send three rows of data, wherein Fig. 1b it needs to send two rows of data to the *summation* function reside in different machines. As a result, if the position of node where the *summation* is done will change, the traffic cost will also change; it is the extra challenge to handle.

*Architecture*: Hadoop has a master node (Job Tracker), and number of slave nodes (Task Trackers) located on remaining nodes. The job tracker handles all the submitted jobs and takes the decision for scheduling and parallelizing the work across the Hadoop cluster. And the task trackers do the work in parallel, allotted by the job tracker. For summating the intermediate data, it needs two things; one piece of code for summation, i.e., *summation phase,* and a manager who will handle the location of that code. The manager resides in job tracker, which has the information where the *summation code* will place for more efficient processing. This architecture will minimize the network traffic in *shuffle* phase.

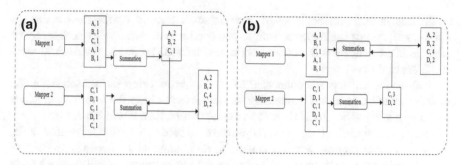

**Fig. 1** MapReduce using summation among different machines

*Summation phase*: In the framework, the *summation phases are* located between *shuffle* phases and *reducer* phases. All the intermediate data acts as input to this and generated output is sent to the *reducer*. It performs the summation of similar *<key, value>* pairs in such a way that each key has single summated pair value. After that, output of *summation phase* along with the similar key has to be delivered to a single *reducer*. In the architecture, the execution of summation is managed by the task tracker at each node. When the manager who is placed in job tracker sends the request to generate the *summation code* at task tracker, task tracker will initialize the instance and specify the tasks attached to the request. Finally, when the task is completed, the *summation code* is removed by task tracker and conveys the message to the manager.

*Manager*: The manager has two main issues—where the *summation code* resides and routes so that the summated intermediated data will generate less amount of traffic.

Summation code placement—Number of *summation codes* will be generated to reduce the traffic along with the path; the path will define from where the intermediate data will go among different machines. To do this, manager has two main questions; by answering that, it will minimize the traffic during *shuffle* phase:

- On which machine the *summation code* will generate for minimum traffic?
- To which machine the intermediate data come from different machines, i.e., what is the route?

For answering these questions, manager needs the whole information about the *map* and *reduce* function along with the positions and the frequency of intermediate data (volume). Furthermore, manager also requires the information about the resources of slave nodes, i.e., the availability of memory and CPU for processing of summation. All these information will be sent by the task tracker to job tracker with the heartbeat. It is sent by task tracker to job tracker so that job tracker has knowledge of whether the task tracker alive or not; here, alive means its working condition. According to that information, the manager will send the info about the *summation code* and the route. The manager has information about the positioning of all the task trackers, so it will send the request to find the central node for summing up the data in different machines by creating small clusters. As the slave nodes get the request, one of them elects itself as the central node and finds whether any of other is interested, finds the central node, and informs the manager. Algorithms 1 and 2 are as follows:

Assume the clusters are connected to each other in a ring form; each node can send information to its next node only so that all nodes have the information and it will create less traffic. But if there is any node failure, then it will bypass it.

Distributed algorithm does not assume the existence of the previous central node; every time according to the requirement, it will change which depends on the frequencies of *<key, value>* pairs and the availability of resources. It will choose a node among a group of different nodes in different machines as a central node.

Assume each node has their own IDs, and the priority of node $Ni$ is $i$, which is calculated according to the availability of resources and the frequency of *<key, value>* pairs, i.e., volume.

Background: Any node $Ni$(among the nodes to which manager sends the function for processing data) tries to find any other active node which is more suitable, by sending a message; if no response comes in $T$ time units, $Ni$ tries to elect itself. Details are as follows:

---

Algorithm 1 for sender node $Ni$ that select suitable node

1. $Ni$ sends an "*Elect Ni*" with $Pi$
2. $Ni$ waits for time $T$
    a. If $Pi < Pj$ it receives "*elect Nj*"
      i. update central node as $Nj$
    b. If no response comes then $Ni$ will be selected as the central node.

---

Algorithm 2 for node receiver $Nj$

1. $Nj$ receive "*elect Ni*"
2. If $Pj > Pi$
    a. Send "*elect Nj*"
3. Else forward "*elect Ni*"

---

# 4 Implementation

The performance baseline is provided by the original scheme (i.e., no summation is provided) and by the proposed method. We create an Oracle VM virtual machine; we configure it with the required parameters (here we use two processors, 3 GB RAM, 20 GB memory) and settings to act as a cluster node (especially the network settings). This referenced virtual machine is then cloned as many times as there will be nodes in the Hadoop cluster. Only a limited set of changes are then needed to finalize the node to be operational (only the hostname and IP address need to be defined). We have created pseudo-cluster. Our prototype has been implemented on Hadoop 2.6.0.0.

In Fig. 2, it gives output sizes of *mappers* on nodes for respective actual sized. It executes the proposed algorithm using the same data source for comparison in Hadoop environment. It shows the data size for different files of size 250 MB, 500 MB, 750 MB, and 1 GB are minimized after applying the *summation* and the file size reduced to 180 MB, 220 MB, 335 MB, and 450 MB, respectively. As a result, the reduction ratios are 28%, 32%, 48%, and 55% for file size of 250 MB, 500 MB, 750 MB, and 1 GB, respectively. The proposed method will work more efficiently as the size increases, and thus it will work well for BigData.

To compute, the capability of the proposed algorithm by comparing traditional hash-based function is shown here. Hash-based partition without *summation*, as default method in Hadoop, makes the traditional hash partitioning for the intermediate data, which are sent to *reducers* without summing up intermediate data. And our proposed method is *summation* within same machine and *summation* among different machines, in which before sending to *reduce* function the intermediate data will be summed up so that it will minimize the size (traffic); sometimes, after summing up, intermediate data at each machine data at *reducer* will be huge because of many *mappers*, so it can be minimized if the summing up will be done among different machines also as per requirements. In Fig. 3, the performance is shown which takes the same file and performs the traditional method, summation on single machine and *summation* among different machines. Here, if the number of keys is increased, the traffic in *shuffle* phase also increases. For example, for 20 keys, traditional, *summa-*

**Fig. 2** Data size at reducer

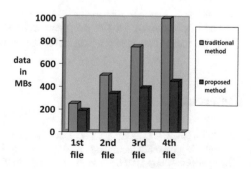

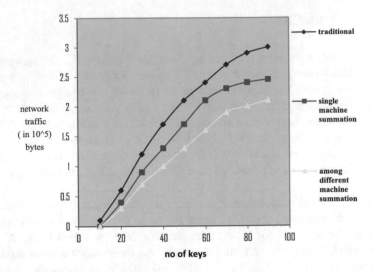

**Fig. 3** Traffic cost

*tion* in single, and *summation* in different machines generate $0.6 * 10^{\wedge}5$, $0.4 * 10^5$, and $0.3 * 10^5$ bytes, respectively.

## 5  Conclusion

The significance of proposed scheme is discussed, i.e., summation in Hadoop MR to process the BigData that minimizes the network traffic produced by intermediate data: intermediate data is output of *map* function. For verification, we have given an architecture where summation functions can be easily attached to the existing MR framework. How the positioning of summation code among various machines will affect the size is also shown, for which we give a distributed election algorithm that fills to find the best suitable positions for the central summation function among various machines. By applying the proposed method, it will reduce the size of data up to 55%.

The implantation for the proposed scheme is in a pseudo-cluster for computing the behavior; it can compute on heterogeneous distributed clusters.

## References

1. Philip Chen, C.L., Zhang, C.-Y.: Data-intensive applications, challenges, techniques and technologies: a survey on big data (2014). Science Direct
2. Apache Haddop HDFS Homepage. http://HADOOP.apache.org/hdfs

3. White, T.: Hadoop: The Definitive Guide, 1st edn. O'Reilly Media (2009)
4. Patel, A.B., Birla, M., Nair, U.: Addressing big data problem using Hadoop and map reduce. IEEE (2013)
5. Ke, H., Li, P., Guo, S., Guo, M.: On traffic-aware partition and aggregation in MapReduce for big data applications. IEEE Trans. Parallel Distrib. Syst. (2015)
6. Blanca, A., Shin, S.W.: Optimizing network usage in MapReduce scheduling (2013)
7. Palanisamy, B., Singh, A., Liu, L., Jain, B.: Purlieus: locality-aware resource allocation for MapReduce in a cloud. ACM (2011)
8. Ibrahim, S., Jin, H., Lu, L., Wu, S., He, B., Qi, L.: Leen: locality/fairness-aware key partitioning for MapReduce in the cloudm. IEEE (2011)
9. Hsueh, S.-C., Lin, M.-Y., Chiu, Y.-C.: A load-balanced MapReduce algorithm for blocking-based entity-resolution with multiple keys (2014)
10. Al-Madi, N., Aljarah, I., Ludwig, S.A.: Parallel glowworm swarm optimization clustering algorithm based on MapReduce. In: 2014 IEEE Symposium on Swarm Intelligence (2014)
11. Liu, L., Han, Z.: Multi-block ADMM for Bigdata optimization in smart grid. IEEE (2015)
12. Liu, Y., Du, J.: Parameter optimization of the SVM for Bigdata. In: 2015 8th International Symposium on Computational Intelligence and Design (ISCID) (2015)
13. Bhushan, M., Singh, M., Yadav, S.K.: Bigdata query optimization by using locality sensitive bloom filter. IJCT (2015)
14. Ramaprasath, A., Srinivasan, A., Lung, C.-H.: Performance optimization of Bigdata in mobile networks. In: 2015 IEEE 28th Canadian Conference on Electrical and Computer Engineering (CCECE) (2015)
15. Ramprasath, A., Hariharan, K., Srinivasan, A.: Cache coherency algorithm to optimize bandwidth in mobile networks. Lecture Notes in Electrical Engineering, Networks and Communications. Springer Verlag (2014)
16. Yildirim, E., Arslan, E., Kim, J., Kosar, T.: Application-level optimization of Bigdata transfers through pipelining, parallelism and concurrency. In: IEEE Transactions on Cloud Computing (2016)
17. Jena, B., Gourisaria, M.K., Rautaray, S.S., Pandey, M.: A survey work on optimization techniques utilizing map reduce framework in Hadoop cluster. Int. J. Intell. Syst. Appl. (2017)

# Secret Image Sharing Over Cloud Using One-Dimensional Chaotic Map

Priyamwada Sharma and Vedant Sharma

## 1 Introduction

Cloud computing has consistently enabled the infrastructure providers to move very rapidly toward new facilities. Such movement has attracted the world toward "pay as you go" model defined by cloud service providers. Almost every task that a user wants to perform is now available over cloud. Such an advancement in cloud computing has made organizations and professionals to migrate their business models and tasks over cloud. However, such swift growth of data toward cloud computing has been raising security and privacy concerns. The users can send their data over cloud for processing and storage without any difficulty at their end, but how secure is the data to be stored over cloud? Such questions are raising a major difficulty in the realization of cloud computing and it is needed to be considered solemnly.

Cryptography is the science of converting plain text to another form which is hard to understand [1]. Communication of encrypted messages such that only willful recipient can decrypt is the main objective of cryptography. Cryptography involves communicating parties, each with a secret key shared between them. Complex computations are used to attain encryption in such a manner that only the anticipated recipient can decrypt the data. Data can be anything and especially in today's world; most of the data is shared in the form of digital images. Hence, when storing the images over untrusted third parties, there is a need arised to secure them from unauthorized access and modifications. One way to secure digital images is image encryption. The secret image is encoded using complex mathematical formulations such

P. Sharma (✉)
School of Information Technology, Rajiv Gandhi Proudyogiki Vishwavidyalaya, Bhopal, India
e-mail: priyamwada14@gmail.com

V. Sharma
University Institute of Technology, Rajiv Gandhi Proudyogiki Vishwavidyalaya, Bhopal, India
e-mail: vedant1998@gmail.com

© Springer Nature Singapore Pte Ltd. 2019
R. K. Shukla et al. (eds.), *Data, Engineering and Applications*,
https://doi.org/10.1007/978-981-13-6351-1_2

that a distorted random-like structure can be obtained in the form of an encrypted image.

On the other hand, visual cryptography is emerging as a new way to secure the images where shares of the plain secret image are generated. Secret image is recovered when all the shares are stacked together. Both image encryption and visual cryptography sound quite the same, but they are different in the sense they realize security to the image. Partitioning of the plain secret image into multiple shares such that no image information is revealed from any of the shares is the primary objective of visual cryptography [2]. The shares can then be transmitted through any untrusted medium. At the receiving side, all the shares are stacked and plain secret image is reconstructed. Introduced by Naor and Shamir [3], visual cryptography was proposed as an easy and feasible approach for secretly sharing the images without needing any difficult cryptographic operations. The concept of visual cryptography can be used to securely store the secret image over untrusted cloud infrastructure. Secret image shares can be generated such that no information can be revealed and stored over cloud.

Simple illustration of visual cryptography is shown in Fig. 1 in which the plain secret image is divided into n shares. After the concept of visual cryptography came to light, several schemes have been proposed. In order to achieve more security of image shares, Shankar and Eswaran [4] proposed a scheme where individual matrices of red, green, and blue channels of the secret image are created by considering the true color band RGB of the secret image pixels. Every value of R, G, and B channels is minimized to further create the submatrices. Division operation is used for the purpose and floor and ceiling functions are used to create two matrices from a single one. In [5], the authors proposed a scheme that uses cellular automata (CA) for visual cryptography. Cellular automaton is a model constituting parallel entities that cooperate with each other and manipulate their progress in a discrete way. If cellular automaton is to be represented in a way that each entity has the state of either 0 or 1, such cellular automata are known as binary cellular automata. A set of rules is used to define state of the cells of cellular automata and acts as secret key for encryption as well as decryption of the plain secret image. Visual cryptography is considered amongst one of the modern cryptographic techniques to share plain secret image. However, the received image size is doubled or more than the original secret image which usually limits the technique.

Additional works other than visual cryptography include share generation using certain polynomial function or any other medium. A secret sharing algorithm to securely transmit the plain image was proposed in [6]. The scheme reduces size of the image shares while maintaining their security. Shamir's secret sharing polynomial was used with pixel values as random coefficients. The scheme was good to share a secret image while reducing amount of traffic on the network. However, it was claimed in [7] that security of the scheme [6] has a limitation of dependency on the permutation stage. In [7], the authors presented a scheme to overcome the limitation of permutation stage. But the scheme [7] still has a limitation of revealing secret image in initial shares. Therefore, further improvements are presented in [8] by repeatedly modifying the share numbers using modulo prime function.

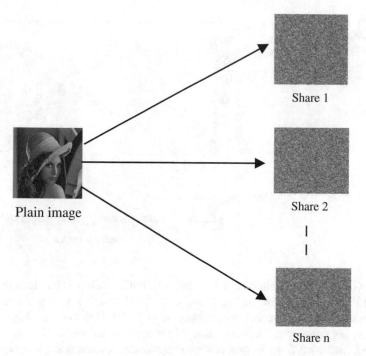

**Fig. 1** Visual cryptography

In this paper, a simple and efficient secret image sharing scheme is presented such that a plain secret image can be stored over cloud infrastructures. The proposed scheme uses one-dimensional chaotic logistic map to generate image shares which can be further stored on the cloud. Rest of the paper is organized as follows: chaotic logistic map and bifurcation of one-dimensional chaotic logistic map are described in Sect. 2. Proposed scheme is discussed in detail in Sect. 3. Section 4 presents results and security analysis. Finally, Sect. 5 concludes the paper.

## 2   Chaotic Logistic Map

Chaotic maps are used as a new way of cryptography. Due to essential properties such as unpredictability, ergodicity, and sensitivity to primary conditions, chaotic systems are emerging very rapidly in building the cryptosystems [9]. A one-dimensional chaotic logistic map is defined as follows:

$$x_n = \mu * x_{n-1} * (1 - x_{n-1}) \tag{1}$$

**Fig. 2** Bifurcation of
one-dimensional chaotic
logistic map

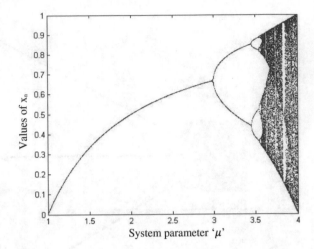

where $\mu$ is the system parameter. Bifurcation of one-dimensional chaotic logistic map depicted in Eq. (1) is shown in Fig. 2, and it can be observed from Fig. 2 that chaotic map possess high random behavior around $\mu = 3.999$. Due to such a high random behavior, we have used 3.999 as value of the system parameter. Chaotic systems require initial conditions to generate random sequences and a slight modification in initial values or system parameter leads a major change in the generated chaotic sequences. Due to such a high random behavior of chaotic maps, a number of chaos-based image encryption schemes have been proposed during the last decade [9–16].

## 3  Proposed Scheme

In the proposed scheme, secret image is divided into two shares using chaotic keystream derived from one-dimensional chaotic logistic map. Two main phases of the proposed scheme are described as follows:

### 3.1  Computation of Initial Value of Chaotic Map Using Secret Key

At the sender's end, 132-bit secret key is used to compute the initial value of the chaotic map. The steps are as follows:

(1)  Let hexadecimal representation of the secret key is
     $k = k_0, k_1, k_2 \ldots k_{32}$
     where $k$ represents the hexadecimal value.
(2)  Now, convert the key into binary form as

$b = b_0, b_1, b_2 \ldots b_{131}$

where $b$ represents the binary value.

(3) Calculate the value $x_{01}$ as

$$x_{01} = b_0 \times 2^{131} + B_1 \times 2^{130} + B_2 \times 2^{129} \ldots + B_{131} \times 2^0$$

(4) Now evaluate the initial value $x_1$ as

$$x_1 = (x_{01}/2^{132}) \, mod \, 1$$

This initial value is used to execute one-dimensional chaotic logistic map depicted in Eq. (1).

## 3.2 Image Encryption and Transmission of Shares to the Cloud

Here, we assume that the two cloud infrastructures are unaware of each other such that there should be no possibility of collision attack. Once initial values of the chaotic map are computed, secret image can be encrypted to form the shares such that the shares can be transmitted for storage on cloud. First of all, execute the chaotic map to get key matrix as follows:

$$initialize \ n=1;$$
$$while \ (true)$$
$$x_{n+1} = \lceil 3.999 \times x_n \times (1 - x_n) \times g \rceil$$
$$n = n + 1$$
$$end$$

where $g$ is the highest gray-level value of the secret image and $n$ represents the chaotic sequence number. A chaotic key matrix $x$ of the same size as of the secret image is obtained when the complete loop is executed. After executing the chaotic map for *rows* * *columns* times, perform the following:

$$initialize \ m=1;$$
$$while \ (true)$$
$$e_m = e_m \oplus x_m$$
$$m = m + 1$$
$$end$$

The resulting image $e$ and chaotic key matrix $x$ can be transmitted to the cloud now [18]. In other words, it can be said that $e$ and $x$ are the shares of secret image and without knowledge of both the shares, secret image cannot be recovered. No information about the secret image is revealed and can be clearly observed from Fig. 3. Complete workflow of the proposed scheme is shown in Fig. 4.

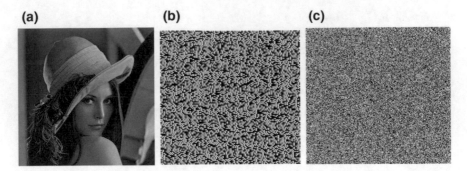

**Fig. 3 a** shows the plain image, **b** and **c** show the non-revealing image shares generated by the proposed scheme

**Fig. 4** Block diagram of the proposed scheme

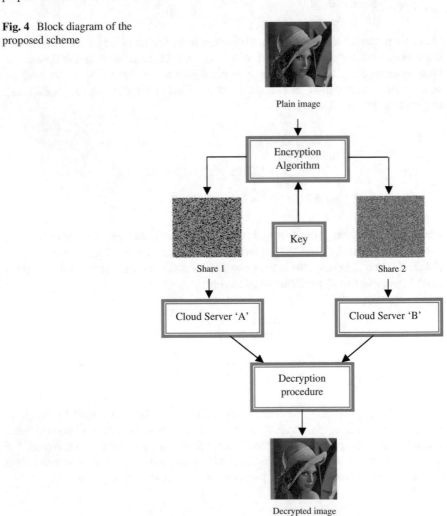

# 4  Experimental Results

In this section, we can simulate proposed scheme in Matlab 2012b. This section can be categorized into number of subsections such as key space analysis, key sensitivity analysis, Histogram analysis, and correlation coefficient analysis. After simulation, we can compare the result of proposed scheme with existing scheme in Table 1 and Fig. 7.

## 4.1  Key Space Analysis

Sensitivity to the secret key is very important for a good image encryption algorithm, and the key space should be large enough to make brute force attacks infeasible. Key analysis for the proposed image cipher has been performed and carried out with results summarized as follows: The proposed image cipher has $2^{132}$-bit secret key. An image cipher with such enormous key space is adequate for resisting all kinds of brute force attacks.

## 4.2  Key Sensitivity Analysis

For an ideal image cipher, a slight change in secret key should produce a completely different encrypted image. In order to perform key sensitivity test, we encrypted secret image with secret key $k_1$ as "67D5EA180B4CF3FA3BF4AD1E27CF2 D7B" and then again we encrypted the same secret image with a slight change in secret key as $k_2$ "67D5E**D**180B4CF3FA3BF4AD1E27CF2D7B". The results are shown in Fig. 5.

**Table 1** Correlation coefficients of two adjacent pixels of the plain image "Lena" and corresponding encrypted shares

| | | Encrypted shares by the proposed scheme | | Encrypted image by [17] |
|---|---|---|---|---|
| | | *Share 1* | *Share 2* | |
| | Horizontal | 0.0137 | 0.0098 | −0.01516 |
| | Vertical | −0.0189 | −0.0114 | 0.01396 |
| | Diagonal | 0.0109 | 0.0134 | 0.02180 |

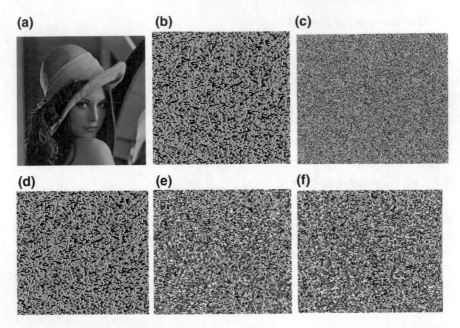

**Fig. 5** Key sensitivity analysis: **a** plain image; **b–c** shares generated by key $k_1$; **d–e** shares generated by key $k_2$; **f** decrypted image using modified key

## 4.3 Histogram Analysis

Histogram is the graphical representation of an image. It exemplifies how pixels in an image are distributed by plotting the number of pixels at each color intensity level. For good image cipher, histogram should be uniformly distributed such that no information about the secret image can be revealed spatially. Histogram analysis of the plain and encrypted shares by the proposed scheme is shown in Fig. 6, and it can be clearly observed that histogram of the image shares is fairly uniform and significantly different from the respective histogram of the plain image.

## 4.4 Correlation Coefficient Analysis

Correlation between adjacent pixels is usually high in the plain image Fig. 7. On the contrary, correlation between adjacent pixels should be as low as possible for the encrypted image. Horizontal, vertical, and diagonal adjacent pixels are considered for computing the correlation coefficients. The correlation coefficient of two adjacent pixels is calculated as:

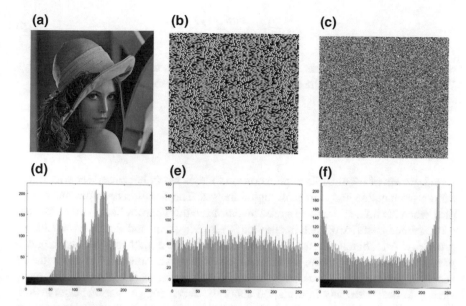

**Fig. 6** Histogram analysis: **a** shows the plain image "Lena". **b–c** show the corresponding image shares. **d–f** show histograms of images shown in **a–c,** respectively

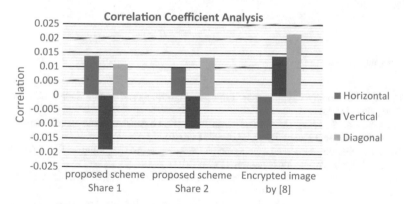

**Fig. 7** Correlation coefficient analysis

$$E(x) = \frac{1}{N} \sum_{i=1}^{N} x_i \tag{2}$$

$$D(x) = \frac{1}{N} \sum_{i=1}^{N} (x_i - E(x))^2 \tag{3}$$

$$cov(x, y) = \frac{1}{N} \sum_{i=1}^{N} (x_i - E(x))(y_i - E(y)) \tag{4}$$

$$r_{xy} = \frac{cov(x, y)}{\sqrt{D(x)} \times \sqrt{D(y)}} \qquad (5)$$

where $x$ and $y$ are the values of two neighboring pixels of the image and, $cov(x, y)$, $D(x)$ and $E(x)$ designates the covariance, variance, and mean.

## 5 Conclusion

Rapid growth of multimedia applications in today's world has raised security issues for communication and storage of digital images. The situation becomes more complex when the images are to be stored on untrusted third parties like cloud infrastructures. Hence, security of such image data is very important and should be considered primarily. A technique to securely store and transmit digital images over cloud is presented in this paper. One-dimensional chaotic logistic map is used to divide the secret image into two obfuscated shares. Experimental results showed that image shares do not reveal any information about the secret image. Various tests including keyspace, key sensitivity, histogram analysis, and correlation coefficient analysis have been performed which demonstrates that the proposed scheme is good to resist various attacks.

## References

1. Schneier, B.: Applied Cryptography: Protocols Algorithms and Source Code in C. Wiley, New York, USA (1996)
2. Liao, X., Lai, S., Zhou, Q.: A novel image encryption algorithm based on self-adaptive wave transmission. Sig. Process. **90**(9), 2714–2722 (2010)
3. Naor, M., Shamir, A.: Visual cryptography. In: Advances in Cryptology—EUROCRYPT'94, pp. 1–12. Springer, Berlin/Heidelberg (1995)
4. Shankar, K., Eswaran, P.: Sharing a secret image with encapsulated shares in visual cryptography. Procedia Comput. Sci. **70**, 462–468 (2015)
5. Yampolskiy, R.V., Rebolledo-Mendez, J.D., Hindi, M.M.: Password protected visual cryptography via cellular automaton rule 30. In: Transactions on Data Hiding and Multimedia Security IX, pp. 57–67. Springer, Berlin/Heidelberg (2014)
6. Thien, C.-C., Lin, J.-C.: Secret image sharing. Comput. & Graph. **26**(5), 765–770 (2002)
7. Alharthi, S., Atrey, P.K.: An improved scheme for secret image sharing. In: IEEE International Conference on Multimedia and Expo (ICME), pp. 1661–1666 (2010)
8. Alharthi, S.S., Atrey, P.K.: Further improvements on secret image sharing scheme. In: Proceedings of the 2nd ACM Workshop on Multimedia in Forensics, Security and Intelligence, pp. 53–58 (2010)
9. Yoon, J.W., Kim, H.: An image encryption scheme with a pseudorandom permutation based on chaotic maps. Commun. Nonlinear Sci. Numer. Simul. https://doi.org/10.1016/j.cnsns.2010.01.041 (2010)
10. Tong, X., Cui, M.: Image encryption with compound chaotic sequence cipher shifting dynamically. Image Vis. Comput. **26**, 843–850 (2008)

11. Behnia, S., Akhshani, A., Mahmodi, H., Akhavan, A.: A novel algorithm for image encryption based on mixture of chaotic maps. Chaos, Solitons Fractals **35**, 408–419 (2008)
12. Pareek, N.K., Patidar, V., Sud, K.K.: Image encryption using chaotic logistic map. Image Vis. Comput. **24**, 926–934 (2006)
13. Patidar, V., Pareek, N.K., Sud, K.K.: Modified substitution–diffusion image cipher using chaotic standard and logistic maps. Commun. Nonlinear Sci. Numer. Simul. **15**, 2755–2765 (2010)
14. Jolfaei, A., Mirghadri, A.: Image encryption using chaos and block cipher. Comput. Inf. Sci. **4**(1), 172–185 (2011)
15. Guarnong, C., Mao, Y., Chui, C.K.: A symmetric image encryption scheme based on 3D chaotic cat maps. Chaos, Solitons Fractals **21**(3), 749–761 (2004)
16. Chen, G.R., Mayo, Y.B., et. al.: A symmetric image encryption scheme based on 3D chaotic cat maps. Chaos Solitins & Fractals **21**, 749–761 (2004)
17. Abdo, A.A., Lian, S., Ismail, I.A., Amin, M., Diab, H.: A cryptosystem based on elementary cellular automata. Commun. Nonlinear Sci. Numer. Simul. **18**(1), 136–147 (2013)
18. Xiang, T., Hu, J., Sun, J.: Outsourcing chaotic selective image encryption to the cloud with steganography. Digit. Signal Process. **43**, 28–37(2015)

# Design and Development of a Cloud-Based Electronic Medical Records (EMR) System

Victoria Samuel, Adewole Adewumi, Benjamin Dada, Nicholas Omoregbe, Sanjay Misra and Modupe Odusami

## 1 Introduction

Movement of bulky files down a hospital corridor is gradually becoming a thing of the past in the developed and developing countries of the world. Electronic Health Record (EHR) systems have been built to solve this problem to an extent. The adoption rate is slow, however, due largely to the low level of awareness on the standards of Health Level 7 (HL7) for the creation of interfaces to communicate with other software. Electronic Medical Records (EMR) can be seen as a streamlined version of an Electronic Health Record (EHR) system. Change is constant in the health care sector and recent advances in technology are piloting this change. The application of technology in healthcare can help to; increase access to healthcare; make patient records more accessible; increase professional communication; create global health networking; and improve efficiency [1]. One Veteran Health Administration research rates its electronic medical record system can improve overall efficiency by 6% per year, and the monthly cost of an EMR may (depending on the cost of the EMR) be offset by the cost of only a few "unnecessary" tests or admissions [2].

Implementing and hosting electronic health records or electronic medical records using cloud technology can help to improve efficiency of medical centers especially in the Nigerian context where paper records are more often used than electronic records. It can offer benefits such as the ability to monitor data over time, recognize patients who are due for preventive visits and screening, watch how patients measure up to certain parameters such as blood pressure reading and vaccinations, and improve the overall quality of delivery in the Nigerian health care causing less delay in information retrieval [3]. Presently, many medical centers move patients' files manually, which is susceptible to destruction and/or defacement of very important pieces of information that can be used for analysis [4, 5]. Furthermore, due to the

V. Samuel · A. Adewumi (✉) · B. Dada · N. Omoregbe · S. Misra · M. Odusami
Center of ICT/ICE Research, CUCRID Building, Covenant University, Ota, Nigeria
e-mail: wole.adewumi@covenantuniversity.edu.ng

© Springer Nature Singapore Pte Ltd. 2019
R. K. Shukla et al. (eds.), *Data, Engineering and Applications*,
https://doi.org/10.1007/978-981-13-6351-1_3

large number of patronage these medical centers receive, the number of files could get bulky which could lead to loss-in-transit [6]. Storage space is also becoming a source of concern as many health centers have been around for some time, and have amassed a large number of records. Lastly, the unthinkable amount of time taken to dig up an old patient's record for processing has in time past led to the loss of lives [7]. The aim of this work, therefore, is to address the aforementioned challenges by designing and developing a cloud-based EMR system. The rest of this paper is structured as follows: Sect. 2 discusses related works while Sect. 3 describes the methodology adopted in developing the proposed system. In Sect. 3.1, the proposed system is implemented as a web application and deployed to a cloud-hosting platform. Section 4 analyses the results obtained while Sect. 5 concludes the paper.

## 2  Related Work

Existing studies in this area can be classified into studies proposing models for cloud-based EHR, studies proposing systems, as well as existing systems.

### 2.1  Studies Proposing Models for Cloud-Based EHR

The study in [8] proposed cloud-based solutions for two separate implementation layouts of EHR systems. The first is for a large hospital while the other is for primary care centers through a network. Similarly, the study in [9] presents the architecture for cloud-based health information systems targeted at two departments of a hospital—the Pediatrics and Obstetrics as well as the Gynecology departments. In addition, the study in [10] proposes a data exchange mechanism that emphasizes interoperability between HISs and e-health applications across a cloud-based service. The studies in this category mostly provide design models and do not necessarily provide any form of implementation.

### 2.2  Proposed Systems

A number of proposed systems have been put forth in various studies. We refer to them as proposed given the fact that they have not been widely adopted for use in hospitals. These studies often adopt cloud technology for a number of reasons. For instance, the study in [11] adopted cloud technology so as to address interoperability issues in EHR systems. The results from the study showed that the cloud-based approach offered significant advantages over the typical client–server architecture used in implementing EHR systems. The studies in [12] and [13] sought to address issues bordering on EHR privacy in the cloud by allowing for efficient storing and sharing of personal health records while also providing a systematic access control mechanism. The study in [14] introduced palm vein pattern recognition-based medical record retrieval system using cloud computing in order to manage emergency scenarios where patients are mentally affected, disabled or unconscious. The study in [15] specifically presents a cloud-based electronic medical record system for maternal and child health outpatient setting.

## 2.3 Existing Systems

OpenEMR [16] is a free and open-source EHR and medical practice management application. It has features such as patient demographics, patient scheduling, prescriptions, billing, and patient portal. It has the goal of providing community support and access to medical records. The system CareMedica introduced in [17] is a web-based electronic health records system that enables a platform for other medical professionals and doctors to collaborate for effective patient care. It gives the loop between the diversity of information needed for optimized diagnoses and the creation of treatment plans. Though it is free and easy to set up, it is not cloud-based making it platform-dependent. FASMICRO introduced in [18] offers reliability, efficiency, and ease-of-use and is aimed at managing records and making them accessible through mobile devices. However, it is not cloud-based. Allscripts Wand introduced the native iPad application for professional electronic health records system and enterprise, and enables a complete approach to how clinicians practice medicine. It gives the mobile healthcare professional the flexibility to move between their desktop and iPad for patient consultations and management [19]. The study in [20] introduced fully integrated modules that can operate on multiple platforms and offer both web and client–server models while [21] designed an Enterprise Electronic Cloud-Based Electronic Health Records System (E2CHRS). The system was designed, implemented, and tested for recording, retrieving, archiving, and updating of patients and other medical records such as electronic medical records, digital medical imaging, pharmacy records, and doctor's notes are all consolidated and accessible in real time. Data analytics on the E2CHRS cloud structured data was proposed to lead to better treatment options, optimal insurance programs, and the possibilities of truly personalized healthcare.

## 3 Methodology

The proposed EMR system was designed using the Unified Modeling Language. Figure 1 depicts the deployment diagram for the proposed EMR. It consists of a layered architectural pattern that has three layers—the user, workstation, and server. The user represents the persons that will interact with the system through the web browser. The workstation consists of the client, which is the web browser with which

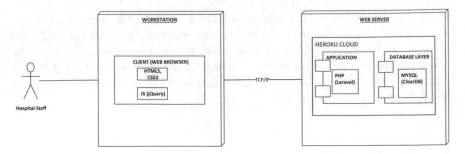

**Fig. 1** Deployment diagram of the cloud-based EMR System

the user has access to the system. HTML5, CSS3, and JavaScript are the languages used for the client side.

The server layer refers to the layer from which the EMR system can be deployed as a web application. It houses the Heroku cloud (which comprises of the application itself and the database layer) where the application was hosted [22]. Apache web server was used to host the application while MySQL was used as a database [23].

## 3.1 System Implementation

This section describes the implementation of the EMR System, which was developed using PHP as the server-side scripting language and Git Bash as the development tool [24, 25]. It comprises of both the application and the cloud service.

Upon authentication and authorization of a user, a dashboard is being loaded from which the user can easily navigate, in order to carry out the duties of his/her department/role. The interface in Fig. 2 represents the admin dashboard. The system essentially offers role-based access to medical records. The admin is the superuser that is able to view all registered users in the hospital.

Some other views that can be accessed include the front desk view of patients, which is accessible to hospital receptionists. This view allows new patient records to be added, edited, or searched by the desk front officers. In addition, the doctor view allows a physician to view a patient's vital signs, enter consultation notes, request lab tests, and prescribe drugs after consultation.

## 4 Discussion

A number of existing Electronic Medical Records have been mostly implemented as web applications [26]. This project, thriving on the recommendation of earlier researchers have decided to take it into the cloud, hence making it easily accessible

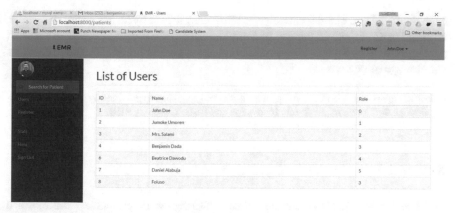

**Fig. 2** Interface showing users' role-based access to the system

at once to all the hospital staffs, it would help improve scalability of the EMR Solution in the event of an increase in number of users (hospital staff) [27].

Unlike the studies in [8–10] that only propose models for cloud-based electronic medical records, the present study also goes ahead to demonstrate design models through a proof-of-concept prototype. This present study employs role-based access to medical records as against the use of biometrics in [14]. Furthermore, this present study is developed as a web-based application that can be hosted on different cloud platforms making it platform-independent unlike the study in [17]. In addition, the hosting company manages the entire infrastructure necessary to deploy such a solution thus allowing the hospital to focus on their primary goal, which is to deliver quality healthcare services to patients. The project has been built with some of the latest technologies, hence, it enjoys the security benefits that come with those time-tested technologies [28].

## 5 Conclusion

This study has both designed and implemented an electronic medical record using cloud technology. Key among the contributions of this study is that the system employs role-based access to medical records. In addition, it replaces the existing approaches to keeping medical records in the Nigerian context, which is largely paper-based or does not utilize cloud computing technology. The application of technology in the medical and healthcare service sectors is constantly increasing and practically unending. Deploying an Electronic Medical Record System to meet a specific hospital's need might be daunting, but the use of agile methodologies can help the hospital quickly adapt to changing circumstances in its environment. The use of a cloud-based EMR would lead to easy accessibility, better patient management, and reduced waiting time at healthcare service delivery points. This project has been deployed as a cloud application to allow easy access by hospital staff. However, with the rapid growth in the use of mobile devices, it would be profitable for such platforms to be exploited. This would bring about greater portability and access to a wider range of persons.

**Acknowledgements** We acknowledge the support and sponsorship provided by Covenant University through the Centre for Research, Innovation and Discovery (CUCRID).

# References

1. Kabene, S.M.: Healthcare and the Effect of Technology: Developments, Challenges and Advancements, IGI Global, Hershey, PA (2010)
2. Evans, D.C., Nichol, W.P., Perlin, J.B.: Effect of the implementation of an enterprise-wide electronic health record on productivity in the Veterans health administration. Health Econ. Policy Law **1**, 163–169 (2006)
3. Rotich, J.K., Hannan, T.J., Smith, F.E., Bii, J., Odero, W.W., Vu, N., Mamlin, B.W., Mamlin, J.J., Einterz, R.M., Tierney, W.M.: Installing and implementing a computer-based patient record system in sub-Saharan Africa: the mosoriot medical record system. J. Am. Med. Inform. Assoc. **10**, 295–303 (2003)
4. Abdulkadir, A.: Medical record system in Nigeria: observations from multicentre auditing of radiographic requests and patients' information documentation practices. J. Med. Med. Sci. **2**, 854–858 (2011)
5. Hamilton, B.E., Martin, J.A., Osterman, M.J.K., Curtin, S.C.: Births: Preliminary data for 2014. National Vital Statistics Reports, 63 (2015)
6. Green, M.A., Bowie, M.J.: Essentials of Health Information Principle and Practices, 2nd edn. Cengage Learning, Delmar (2011)
7. Thompson, C.D.: Benefits and risks of Electronic Medical Record (EMR): An interpretive analysis of healthcare consumers' perceptions of an evolving health information systems technology, UMI Dissertations Publishing. ProQuest, Pennsylvania (2013)
8. Fernández-Cardeñosa, G., de la Torre-Díez, I., López-Coronado, M., Rodrigues, J.J.: Analysis of cloud-based solutions on EHRs systems in different scenarios. J. Med. Syst. **36**, 3777–3782 (2012)
9. Lupşe. O.S., Vida, M.M., Tivadar, L.: Cloud computing and interoperability in healthcare information systems. In: The First International Conference on Intelligent Systems and Applications, pp. 81–85 (2012)
10. Radwan, A.S., Abdel-Hamid, A.A., Hanafy, Y.: Cloud-based service for secure electronic medical record exchange. In: 22nd International Conference on Computer Theory and Applications (ICCTA), pp. 94–103 (2012)
11. Bahga, A., Madisetti, V.K.: A cloud-based approach for interoperable electronic health records (EHRs). IEEE J. Biomed. Health Inf. **17**, 894–906 (2013)
12. Xhafa, F., Li, J., Zhao, G., Li, J., Chen, X., Wong, D.S.: Designing cloud-based electronic health record system with attribute-based encryption. Multimedia Tools Appl. **74**, 3441–3458 (2015)
13. Wu, R., Ahn, G.H., Hu, H.: Secure sharing of electronic health records in clouds. In: 8th International Conference on Collaborative Computing: Networking, Applications and Worksharing (CollaborateCom), pp. 711–718 (2012)
14. Karthikeyan, N., Sukanesh, R.: Cloud based emergency health care information service in India. J. Med. Syst. **36**, 4031–4036 (2012)
15. Haskew, J., Rø, G., Saito, K., Turner, K., Odhiambo, G., Wamae, A., Sharif, S., Sugishita, T.: Implementation of a cloud-based electronic medical record for maternal and child health in rural Kenya. Int. J. Med. Inform. **84**, 349–354 (2015)

16. OpenEMR community Home.: OpenEMR Website. Retrieved from OpenEMR Website: http://www.open-emr.org/ (2016)
17. CareMedica Systems.: Retrieved from CareMedica Systems: http://www.caremedica.com.ng/caremedica-how-it-works.php (2016)
18. FASMICRO.: Retrieved from FASMICRO, http://fasmicro.com/emr (2014)
19. http://www.allscripts.com/en/solutions/ambulatory-add-ons/wand.html
20. https://www.eclinicalworks.com/products-services/eclinicalworks-v10-ehr-suite/
21. Abayomi-Alli, A.A., Ikuomola, A.J., Robert, I.S., Abayomi-Alli, O.O.: An enterprise cloud-based electronic health records system-E2CHRS. J. Comput. Sci. Inf. Technol. (JCSIT) 2(3), 21–36 (2014)
22. Heroku.: About: What is Heroku? Retrieved from Heroku Website: https://www.heroku.com/what (2016)
23. Sparx Systems.: UML 2 Deployment Diagram. Retrieved from Sparx Systems Web Site: http://www.sparxsystems.com/resources/uml2_tutorial/uml2_deploymentdiagram.html (2016)
24. Finley, K.: The Problem With Putting All the World's Code in GitHub. Retrieved from Wired: http://www.wired.com/2015/06/problem-putting-worlds-code-github/ (2016)
25. Williams, A.: GitHub Pours Energies into Enterprise—Raises $100 Million From Power VC Andreessen Horowitz. Retrieved from TechCrunch Website: http://techcrunch.com/2012/07/09/github-pours-energies-into-enterprise-raises-100-million-from-power-vc-andreesen-horowitz/ (2016)
26. Fraser, H., Biondich, P., Moodley, D., Choi, S., Mamlin, B., Szolovits, P.: Implementing electronic medical record systems in developing countries. J. Innov. Health Inform. 13, 83–95 (2005)
27. Armbrust, M., Fox, A., Griffith, R., Joseph, A.D., Katz, R., Konwinski, A., Lee, G., Patterson, D., Rabkin, A., Stoica, I., Zaharia, M.: A view of cloud computing. Commun. ACM 53, 50–58 (2010)
28. Gousios, G., Vasilescu, B., Serebrenik, A., Zaidman, A.: Lean GHTorrent: GitHub data on demand. In: Proceedings of the 11th Working Conference on Mining Software Repositories, pp. 384–387. ACM (2014)

# Log-Based Approach for Security Implementation in Cloud CRM's

Madhur Patidar and Pratosh Bansal

## 1 Introduction

Customers are one of the most valuable assets for an organization. As a result, dependency on customers for an organization and developing customer's trust becomes a key issue for a company's business process. Not only retaining the existing customers but attracting new customers is also vital. So understanding the customer's needs and customer satisfaction plays a crucial role. Hence, for increasing customer's lifetime value with the organization and improving business processes, a proper understanding of the current business trends and keeping a track of customers association with an organization acts as an important metric.

E-commerce is a similar process that deals with buying and selling of products, wherein the transactions are performed through the web which comprises many categories such as B2B, B2C, C2B, C2C, etc. Business-to-Consumer (B2C) is an important part of e-commerce, wherein the consumer is the final customer. Several techniques are developed so that the shoppers turn into buyers efficiently for an organization. B2C marketing has evolved over the time on account of knowing the audience in a better way, analyzing the online contents generated by users, building brand loyalty, etc. Also, these involving best practices for the promotion of products and services have many CRM solutions built for them.

Customer Relationship Management (CRM) plays a vital role when it comes to keeping track of a company's interactions with its customers and activities such as emails, website accessed, etc. It is mainly used for providing better services to the customers and helps the sales teams to work effectively and focused. One of the main benefits of CRM applications is that data from heterogeneous departments are

M. Patidar (✉) · P. Bansal
Department of Information Technology, IET, Devi Ahilya Vishwavidyalaya, Indore, MP, India
e-mail: madhurpatidar15@gmail.com

P. Bansal
e-mail: pratosh@hotmail.com

© Springer Nature Singapore Pte Ltd. 2019
R. K. Shukla et al. (eds.), *Data, Engineering and Applications*,
https://doi.org/10.1007/978-981-13-6351-1_4

stored centrally and can be accessed anytime whenever required. Improved relationships with the existing clients help ensure better sales, identification of customer requirements in an effective and planned manner, and also determining which of the customers are profitable. Satisfaction of customers is focused upon largely. CRM applications have been one of the crucial factors for an organization's motivation toward moving over cloud and one of the leading application hosting investments [1].

Cloud-based CRM's are becoming increasingly popular today on account of several benefits over the traditional CRM's as discussed in Sect. 3.1 ahead. Being a cloud application, these are also susceptible to attacks and other security concerns as discussed in Sect. 3.2 ahead. We followed a log-based approach to overcome and minimize these concerns as discussed in Sect. 4 ahead.

The motive behind this paper is to identify certain loopholes with respect to security and privacy over the cloud-based CRM solutions and to design a system that monitors and alerts such undesired issues through a proper log management mechanism.

The paper is structured as follows. The current section gives a brief overview of the paper. The next section subsequently deals with the literature review followed by significance of cloud CRM's in enterprises, the security concerns associated, logs' overview, and approaches for log management. The next section includes the methodology and outcomes followed by the conclusion and future work at the end.

## 2 Literature Review

Works related to identifying security concerns in cloud-based CRM applications and approaches to minimize them have been proposed by various authors. Huei Lee, Kuo Lane Chen, Chen-Chi Shing, and Marn-Ling Shing highlight the security concerns with cloud CRM's in their papers [2]. S. Xiao and Genguo Cheng gave a research based on cloud CRM's [3]. Rajiv R. Bhandari and Nitin Mishra focused on cloud implications in context to CRM applications and implementation of auditing techniques to achieve the level of security in cloud-based applications. At the end, the author also mentions the significance of log management in cloud-based enterprise applications such as CRM's [4]. Raffael Marty in his paper considered log as an important metric for forensic purposes. The author further highlights the necessity for log management in the cloud applications and elaborates various kinds of log records [5]. Mahdi Seify emphasized security through risk management for cloud CRM solutions [6]. Julia Vuong and Simone Braun provided a secured design-based solution for ensuring security for multi-tenant cloud CRM's. Also, the authors provided security algorithms for encryption that focuses on optimizing the performance of the system [7]. Miguel Rodel Felipe, Khin Mi Mi Aung, Xia Ye, and Wen Yonggang proposed a cloud CRM model which is based on full homomorphic encryption of database to provide a secured solution [8]. M. Vanitha and C. Kavitha emphasized secured encryption algorithms for encrypting data at the client and the network ends, helpful for enterprise applications such as CRM's dealing with lots of transactional data [9].

So by studying the above research papers, we inferred that the authors identified the main problems with cloud CRM's as the security concerns which may compromise the sensitive data stored of the users. Also, by the authors' review over the importance of log management in cloud-based enterprise applications, we got a motivation toward adopting a log-based approach for eliminating the security issues concerned with cloud CRM's.

# 3 Research Background

## 3.1 Significance of Cloud CRM's

CRM's can be traditional also known as on-premise or it can be hosted on cloud. These are the two broad categorizations for CRM's.

Traditional CRM's refer to those where the management is done by the in-house IT department with the organization's control over the data. CRM, being sales driven, is based on direct feedback from the customers and is often automated. Although the flexibility for customization is more here, these CRM's usually have higher upfront costs. Also, it is time-consuming in fully integrating it into the business environment of client. So, these are suitable for businesses with more complex needs and technical expertise. Hence, due to its high maintenance and costs, alternative CRM's such as cloud-based CRM's are increasingly becoming popular [10].

Cloud-based solutions provide better services on account of benefits such as IT cost savings, reliability, scalability, availability, and disaster recovery features which ensures business continuity. The management is being looked upon by the service providers, and the SLA conditions lead to a proper delivery and maintenance of the services. Hence, the flexibility provided by cloud solutions encourages more and more business organizations to adopt the same.

Following are some of the statistics which proves that enterprises will adapt cloud-based solutions at a rapid pace in near future.

- By 2018, cloud-based platforms will host more than half of the infrastructure of more than 60% of enterprises [11].
- By 2019, cloud apps will be a part of 90% of the total mobile data traffic which will subsequently be attaining 60% compound annual growth rate (CAGR) [11].

The use of cloud CRM's is increasing day by day, wherein the services are provided through a remote-based location. CRM applications based on cloud have a significant proportion among other cloud applications as per the CSA Survey Report 2016 as shown in Fig. 1.

Some of the key benefits of cloud CRM's can be summarized below:

**Fig. 1** Cloud applications' ratio as per CSA survey report 2016

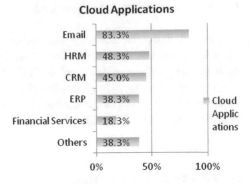

- Based on real-time and updated data of the customers, it saves time involved for uploading, backup, and maintenance purposes.
- Data analysis process became faster and smoother as keeping track of business emails, user's activities including the mails forwarded, and other interactions have greatly enhanced.
- Integration of cloud CRM with big data platforms, which being cloud-friendly, helps in processing huge volumes of data.
- Synchronization with mobile and tablet devices helps in better functioning.
- Focuses on a proper training of the employees by offering free trials for the solutions, documentations, and ongoing support.
- Better organization of data, integration of data from multiple sources, and proper visualization of data which helps in faster decision-making process.
- Cost reduction is one of the main factors as cloud CRM's are based on subscription, thereby eliminating hardware and license purchase costs, installation and maintenance processes, etc.
- Being accessible from anywhere and at any time and with use of additional features such as mobile apps, it increases employee productivity and hence the business growth and optimization.
- Features such as easy deployment help cloud CRM solutions as a preferred one for the small- and medium-sized businesses even with small expertise in technology.

IT environment is greatly affected by the increased adoption of social, mobile, analytics, and cloud (or SMAC). Preference for Software as a Service (or SaaS) by ERP and CRM software companies like Oracle Corp. (ORCL), IBM Corp. (IBM), and Microsoft Corp. (MSFT) forces them to invest billions of dollars in cloud-related products, solutions, and initiatives. Several cloud CRM solutions are available in market today. Salesforce.com is the most popular among them. Figure 2 shows popular cloud CRM vendors.

Hence, we can say that there is a great requirement for maintaining the data associated with cloud CRM's as more and more organizations are getting associated with the same. But, on the other hand, there are some risk factors involved too as discussed in the next section.

**Fig. 2** Cloud CRM software vendors with their proportion in market [12]

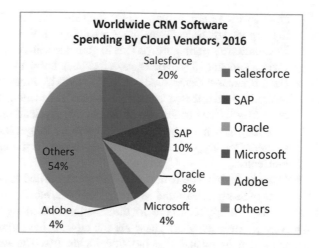

## 3.2 Security Concerns in Cloud CRM's

Security of data is one of the major concerns as far as storing data on the cloud is concerned. CRM applications are no exception to it. Multiple CRM records are maintained by the service providers. So authorization, safety, and access control of CRM data are of utmost importance for an organization. Data breaches can have unpredictable effects on an organization. Today, for various advertising purposes, personal data of customers is being taken through the social media. A single instance for a cloud CRM contains large amount of confidential and proprietary details regarding the customer. Hence, data privacy issues are vital for both corporate and the consumer. Although CRM vendors try to provide measures for safeguarding our data, still vulnerability of data leakage cannot be fully stopped, hence, giving rise to frauds and cybercrimes too.

There are number of security concerns possible with the data stored over cloud CRM's. Some of the main concerns include:

- **Improper activities at network, client, and server ends**: This includes any vulnerable and abnormal activities observed, e.g., servers being compromised, change in the host activities or any other manipulations over the network end, bulky traffic and multiple packets over the network, intrusion attempts, machines not responding, etc.
- **Downtime/outages**: Even short and infrequent CRM outages can affect business to a great level incurring huge losses.
- **DDOS attacks**: These are the attacks that cause the cloud services to be disrupted, thus making it inaccessible. The targeted components may be network bandwidth, disk storage, RAM, etc.
- **Malware attacks**: Security mechanisms must be implemented in order to avoid malicious intends and to save the systems from halts, loss of data, damages caused to the hardware components and the software or the applications, any system inop-

erability issues, etc. An example of such attack is **Zeus Trojan** that has been known to be spotted on Salesforce.com, the leading CRM solutions provider in the world. The malware's main purpose was to gather sensitive business data [13]. Similarly, variant known as **Dyre** or **Dyreza** has been known to target Salesforce.com.

- **Data breaches**: Data stored on cloud CRM's is vulnerable to breaches if the proper designing of databases is not done. Users' data can be targeted for attacks even if there may be flaw in any one of the client's applications. As a result, other client's data may also get exposed. As the users no longer have control over their data, so any of the cryptographic techniques to protect the integrity of their data proves ineffective.
- **Data-in-flight**: The data that is being transmitted is susceptible to cyberattacks, risks of being transmitted to wrong locations, etc. A strong encryption mechanism needs to be applied to protect the sensitive data along its motion.
- **Masquerade attacks**: These are the attacks that focus mainly on obtaining personal data by misusing an individual's identity. The attackers, by taking into trust the customers through false image of a company's representative, can actually sell data to other parties which can further use it for identity theft purposes.
- **Service providers' dependencies**: The users are not sure how is their data being managed by the service provider. As the data can be sensitive and private, issues concern the security of the same arises.
- **Improper designing of SLA's**: Policies regarding deleting the traces of data after the contract, planned maintenances from downtimes, backups, and disaster recoveries might not be properly formulated in the SLA's.

Looking at these attacks, we can infer that people cannot rely exclusively on their SaaS providers for securing data inside their CRM applications.


## 3.3  Logs in Cloud CRM's

Logs refer to the events or activities by users and entities within the networks of the system. Being an important aspect of analytical data, logs are very useful to monitor business processes and faults in infrastructural components and applications. Log records help in identifying error conditions, unauthorized access of resources, patterns of user behavior, incidents alerting, any nonprivileged access, etc. Presently, for collection of logs over cloud CRM's for any forensic investigations, there are some cloud simulation tools available which eliminate dependency on the CSP. Examples of some simulators include CloudSim, Cloud Analyst, Green Cloud, etc. Evaluation and monitoring of logs are very important in context to cloud forensic investigations as privacy of customers' information is directly linked with the business and financial perspectives of an organization [14].

## 3.4   Log Management Approaches

Several approaches helpful in managing logs have been proposed to effectively minimize security concerns over cloud CRM's. Some of the useful approaches are as follows:

- **SIEM**—It refers to security information and event management. This includes user activities monitoring, using log correlation to track malicious events, aggregation of log data, real-time analysis of alerts generated by applications, analysis of log files, reporting through dashboards, etc. [15].
- **SANS logs' reports**—This includes some critical log reports that help in determining the techniques to be followed for log management, e.g., authentication and authorization report which includes login failures and successes by users, suspicious network traffic patterns, etc. [16].
- **SOC**—It refers to security operations center where the main focus is upon detecting, analyzing, and responding to cyber incidents through 24/7 monitoring processes by analyzing the activities across networks, servers, and databases of an organization.

## 4   Proposed Methodology

Enterprise data comprising of customers records, business records, chats, emails, phone records and tracks, last sales records, timestamps, and past records act as an important source for security and IT operations and therefore motivates us for a systematic log management process.

In order to maintain the security principles, the following methods are being focused upon by us:

- **Log monitoring process**: A continuous monitoring process of generated logs in order to detect any suspicious or malicious threats over the privacy and security of data report the same through dashboards (identifying abnormal activities not adhering to standard patterns). To keep a track of the packet flow, web-based activities, the type of request by the client, IP of client, date/time, browser type, and other recorded instances at the server from the requests sent through the browser. For achieving the same, we will follow the approaches as mentioned in Sect. 3.4.
- **Predictive analysis through alerting**: Reporting of security issues analyzed through log monitoring by a well-defined automation of events or alerts.
- **Log mining and aggregation techniques**: Log management measures include data aggregation from various sources such as servers, networks, databases, etc.
- **Post-attack identifications/Forensic analysis**: In case of any suspicious attempts for security breaches, analyzing the causes of attacks through the symptoms observed will help us in preparing and safeguarding our systems from attacks in future.

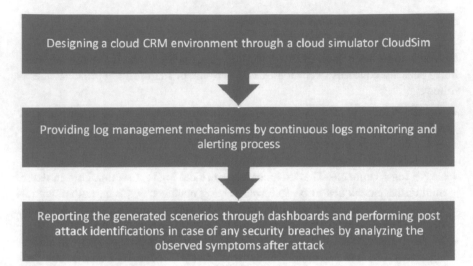

**Fig. 3** Proposed methodology

- **Appropriate troubleshooting**: Based on the log analysis and forensic procedures applied, it will be easier to identify and solve the potential faults and other flaws in the application.
- **Fault tolerance techniques**: In order to ensure 24 * 7 business operations or business continuity, strategies such as backups and fault tolerance techniques to be focused upon are covered here.

The work also focuses on the following concerns with respect to log management.

- Log centralization.
- Log integrity preservation.
- Log records retention and maintenance.

The methodology followed is as shown in Fig. 3 and focuses on analyzing different instances of the log records such as timestamps, logged in user entries, browser used, IP addresses used, reasons for access denials, etc. To achieve the same, we use a cloud simulator CloudSim for creating a cloud environment and generating the logs as shown in Fig. 4.

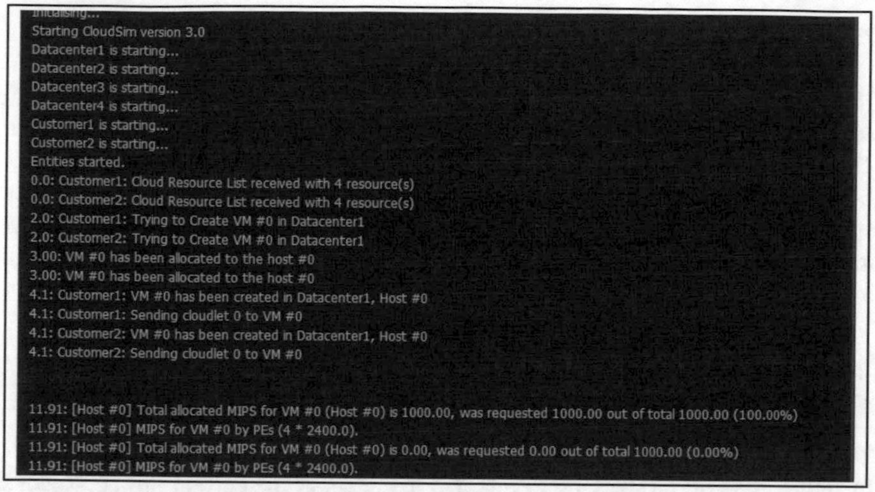

**Fig. 4** Log generation through cloud simulator

## 5 Proposed Outcomes

The resulting outcome will mainly help in retaining the security principles and business continuity aspects with respect to cloud-based CRM firms, and hence will help in achieving the following:

- The faster recovery and troubleshooting processes along with the 24 * 7 continued business operations will help in maintaining a smooth relationship of a company with its customers.
- Log analyses from different domains such as application, network, and interface are quite useful.
- A proper, secure, reliable, and prompt cloud service in combination with systematic log management generation process will result in a fast growth sector particularly for CRM and ERP applications.
- Faster data analysis through dashboards.
- Enhanced customer satisfaction for organizations on account of no issues concerning security and privacy regarding CRM-sensitive data.
- Systematic monitoring process to keep track of activities helps in assisting regarding logs' identification and analysis processes.
- Useful in all web-based domain areas such as organizations involving keeping a track of large number of customer records where customer trust regarding privacy of their data acts as a vital element.
- Useful for system administrators, security analysts, etc. where the job profiles require strict monitoring and auditing processes.
- Helpful in protecting the sensitive data stored in different governmental, commercial, and industrial organizations.

## 6  Conclusion

The rate of enterprises moving toward the cloud suggests the need for systematic security measures and for that purpose log management is a worthy of consideration. The paper provides a roadmap to provide a secured solution by maintaining logs for cloud-based CRM applications. This work mainly concentrates on monitoring and alerting process, troubleshooting, and forensic procedures in case of any abnormalities observed with context to cloud CRM's. The resulting outcomes are quite beneficial for organizations as discussed in Sect. 5.

## 7  Future Work

In future, we will be working to broaden the areas of the current project to cover majority of the areas of time cloud environments. The project can be further utilized for log management in case of large-scale organizations dealing and maintaining big data. Other works will include analyzing the results after performing live attacks, generation of large number of data sets, and working on the security requirements on the same and performing the results on live data centers used for server requirements.

## References

1. Petkovic, I.: CRM in the cloud. In: 2010 8th International Symposium on Intelligent Systems and Informatics (SISY), pp. 365–370. IEEE (2010)
2. Lee, H., Chen, K.L., Shing, C.C., Shing, M.L.: Security issues in customer relationship management systems (crm). In: Decision Sciences Institute 37th Annual Conference Bricktown-Oklahoma City March, vol. 1 (2006)
3. Xiao, S.: Application research of CRM based on SaaS. In: 2010 International Conference on E-Business and E-Government (ICEE), pp. 3090–3092. IEEE (2010)
4. Bhandari, R.R., Mishra, N.: Encrypted IT auditing and log management on cloud computing. Int. J. Comput. Sci. Issues (IJCSI) **8**(5) (2011)
5. Marty, R.: Cloud application logging for forensics. In: Proceedings of the 2011 ACM Symposium on Applied Computing, pp. 178–184. ACM (2011)
6. Seify, M.: New method for risk management in CRM security management. In: Third International Conference on Information Technology: New Generations, 2006, ITNG 2006, pp. 440–445. IEEE (2006)
7. Vuong, J., Braun, S.: Towards efficient and secure data storage in multi-tenant cloud-based CRM solutions. In: 2015 IEEE/ACM 8th International Conference on Utility and Cloud Computing (UCC), pp. 612–617. IEEE (2015)
8. Felipe, M.R., Aung, K.M.M., Ye, X., Yonggang, W.: StealthyCRM: a secure cloud CRM system application that supports fully homomorphic database encryption. In: 2015 International Conference on Cloud Computing Research and Innovation (ICCCRI), pp. 97–105. IEEE (2015)
9. Vanitha, M., Kavitha, C.: Performance enhanced security for enterprise cloud application. In: 2016 International Conference on Computer Communication and Informatics (ICCCI), pp. 1–5. IEEE (2016)

10. Ghafarian, A.: Foreniscs analysis of cloud computing services. In: Science and Information Conference (SAI), 2015, pp. 1335–1339. IEEE (2015)
11. Columbus, L.: Roundup of cloud computing forecasts and market estimates Q3Update. http://www.forbes.com/sites/louiscolumbus/2015/09/27/roundup-of-cloud-computing-forecasts-and-market-estimates-q3-update-2015/#840e1b86c7ad (2015)
12. Stamford, C.: Gartner says customer relationship management software market grew 12.3 percent. http://www.gartner.com/newsroom/id/3329317
13. Osborne, C.: Zeus variant targets Salesforce.com accounts, SaaS applications. http://www.zdnet.com/article/zeus-variant-targets-salesforce-com-accounts-saas-applications
14. Shenk, J.: Log management in the cloud: a comparison of in-house versus cloud-based management of log data. A SANS Whitepaper, October (2008)
15. Balaji, N.: Security information and event management (SIEM)—a detailed explanation. https://gbhackers.com/security-information-and-event-management-siem-a-detailed-explanation/
16. The 6 categories of critical log information. https://www.sans.edu/cyber-research/security-laboratory/article/sixtoplogcategories

# Performance Analysis of Scheduling Algorithms in Apache Hadoop

Ankit Shah and Mamta Padole

# 1 Introduction

Prevailing variety of applications are generating huge amount of data every day, which is much beyond our imagination. Knowingly or unknowingly, we are generating or working with big data. Data these days is no more static in nature, but it is very dynamic. Therefore, the challenge lies not just in processing big data but also in storing, transmitting, and securing big data. Thus, big data application opens a new door for technology and identifies betterment of humanity.

The term big data refers to colossal and complex set of data that cannot be processed in a traditional way [1]. Apache Hadoop [2] is the most suitable open-source ecosystem for processing big data in a distributed manner. Google's MapReduce [3] is the best proposed programming framework for big data processing solution under the umbrella of Hadoop. Hadoop is not just software but it is a framework of tools for processing and analyzing big data.

Big data not only demands faster processing, but it also demands better analysis, security, authenticity, scalability, and more. One of the important aspects of any processing tools is to process it faster. MapReduce satisfies most of the big data processing demands such as scalability, fault tolerance, faster processing, and optimization. However, MapReduce has some limitations while considering performance and efficiency of big data. By considering this fact, many researchers and industries have worked on to overcome the limitation of MapReduce model. The goal of this study is to measure the performance of various scheduling algorithms on different big data applications. The paper discusses big data processing performed using

A. Shah (✉)
Shankersinh Vaghela Bapu Institute of Technology, Gandhinagar, India
e-mail: shah_ankit101@yahoo.co.in

M. Padole
The Maharaja Sayajirao University of Baroda, Vadodara, India
e-mail: mpadole29@rediffmail.com

© Springer Nature Singapore Pte Ltd. 2019
R. K. Shukla et al. (eds.), *Data, Engineering and Applications*,
https://doi.org/10.1007/978-981-13-6351-1_5

Hadoop/MapReduce model. This study aims to identify better scheduling model depending upon big data application that needs to be processed.

## 2 Hadoop Ecosystem

Hadoop is open-source software comprising of framework of tools. These tools provide support for executing big data applications. Hadoop has very simple architecture. Hadoop 2.0 version primarily consists of three components as shown in Fig. 1.

1. HDFS (Hadoop Distributed File System): It provides distributed storage of data over Hadoop environment. It stores data and metadata separately.
2. YARN (Yet Another Resource Negotiator): YARN is responsible for managing the resources of Hadoop cluster.
3. MapReduce: It is the programming model on top of YARN responsible for processing of data in the Hadoop environment. It performs the computation.

## 2.1 HDFS

Hadoop HDFS has master/slave architecture. Master node has two components called resource manager and namenode. Slave on each node of a cluster is having node manager and datanode. Namenode and datanode are under the umbrella of the HDFS, while resource manager and node manager are under the umbrella of YARN.

The big data applications in Hadoop first assign the task to the master node. Master node will distribute the task among multiple slaves to perform computation, and end result will be combined and given back to the master node.

**Fig. 1** Hadoop 2.0 architecture

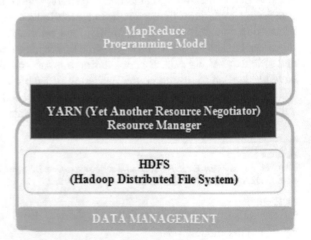

In case of distributed storage, it is important to give indexing for faster and efficient data access. The namenode that resides on the master node contains the index of data that is residing on different datanodes. Whenever an application requires the data, it contacts the namenode that routes the application to the datanode to obtain the data.

Hardware failures are bound to happen, but Hadoop has been developed with efficient failure detection model. Hadoop has de facto fault tolerance support for data. By default, Hadoop maintains three copies of file on different nodes. Therefore, even in case if one datanode fails, system would not stop running as data would be available on one or more different nodes.

Fault tolerance does not handle the failure of just slave nodes, but it also takes care of failure of master node. There is no single point of failure in case of master node. Hadoop maintains multiple copies of namenode on different computers as well as maintains two masters, one as a main master and other as a backup master.

Programmers need not worry about the questions like where the file is located, how to manage failure, how to split computational blocks, how to program for scalability, etc. Hadoop implicitly manages all these efficiently. It is scalable, and its scalability is linear to the processing speed.

In Hadoop 1.x version, MapReduce manages both resources and computation. However, Hadoop 2.x splits the two responsibilities into separate entities by introducing YARN.

## 2.2 YARN

YARN is a framework to develop and/or execute distributed applications. As shown in Fig. 2, components in the YARN-based systems are Global Resource Manager (RM), Application Master (AM) for each application, Node Manager (NM) for each slave node, and an application container for each application running on a node manager.

Resource manager has two main components: Scheduler and application manager. The scheduler schedules the tasks based on availability and requirement of resources. The scheduler schedules the task based on capacity, queues, etc. The scheduler allocates the resources by taking consideration of memory, CPU speed, disk capacity, etc. The application manager accepts the job from client and negotiates to execute the first container of the application. The application manager provides the failover mechanism to restart the services, which might have failed due to application or hardware failure. Each application manager tracks the status of individual application.

## 2.3 MapReduce Programming Model

MapReduce is Google's programming model for processing the colossal amount of data. This model consists of two important phases, i.e., maps and reduces. As

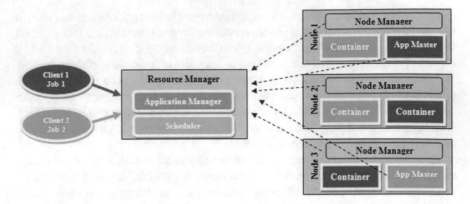

**Fig. 2** YARN architecture

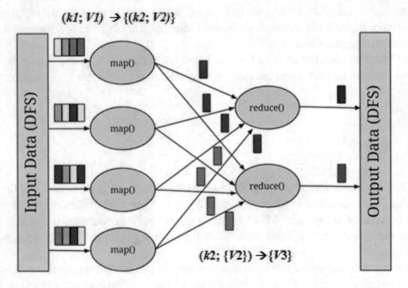

**Fig. 3** MapReduce model [4]

shown in Fig. 3 in "map" phase, it takes input as key–value (k, V) pair and produces intermediate key–value pair (k1, V1) → {(k2, V2)} as a result, while in "reduce" phase, it takes a key and a list of the keys and values and generates the final output as key/value (k2; {V2}) → {V3} pair. In distributed processing, it is important to take consideration of data locality. If data to be processed is located near, then it can reduce the time of transmission and can achieve better performance. MapReduce can use this functionality during MapReduce function. In MapReduce, each map function will take place on local data and output will be stored to temporary storage.

A master node coordinates the input data only after an input is processed. In the next phase, i.e., shuffle phase, it randomly generates values assigned and then sorts it according to the assigned values. Now in reduce phase, it processes the intermediate key–value data and produces the final output.

# 3 Hadoop Schedulers

Hadoop supports three scheduling schemes in MapReduce framework: FIFO, Capacity [5], and Fair [6] scheduler. MapReduce1 (MR1) comes with all three with FIFO as default scheduler, while MR2 comes with capacity and fair scheduler, which can be further configured with delay scheduler to address the locality issue.

## 3.1 Capacity Scheduler

This is the default scheduler, which comes with the MR2 or YARN. The capacity scheduler's configuration supports multiple queues, which can be allocated to multiple users based upon tasks or organization. This scheduler is designed with an idea that same cluster can be rented to multiple organizations and resources may be divided into several users. Thus, the organization can divide their resources across multiple departments or users depending upon their tasks or the cluster can also be divided among multiple subsidiary organizations. Each queue can be configured with fix portion of resources, which can be soft or hard. Generally, resources are soft having elastic allocation but can also be configured for hard approach.

Capacity scheduler makes use of First-In First-Out (FIFO) scheduling if multiple jobs are in the same queue. Suppose a job comes into the queue "A" and if queue "A" is empty, then it allocates all the resources to the first job. This would utilize more resources and then configured capacity of queue, particularly if queue allocation is elastic and job requires more resources. When a new job comes in queue "B", assuming that the first job is still running and using the resources more than its allocated capacity, then tasks of first job will be killed to free up the resources and allocate that resources to second job. Suppose if another job comes to the queue "A" or "B", the capacity scheduler will process it like FIFO or FIFO with priority. There are many features available like capacity guarantee, elasticity, security, etc. that can be customized as per requirement.

## 3.2 Fair Scheduler

Fair schedulers have similar queue configuration as discussed in capacity scheduler. Jobs would be submitted to the queue, which is termed as a "pool" in case of fair scheduler. Each job will use the allocated resources to their pools. As in capacity scheduler, in FIFO approach, the jobs which are coming late have to wait till the time first job finishes or resources made available, so this problem is solved in the fair scheduler that the jobs which have waited in the queue would be picked up and would be processed in parallel with the same amount of resources shared by the applications which are in the same queue. Fair scheduler supports three scheduling policies, that is, FIFO, Fair, and Dominant Resource Fairness (DRF).

In FAIR-FIFO scheduling policy, if multiple jobs are in the same queue, then resources will be allocated to the job, which enters first in the queue, and each job will run serially. However, fair sharing is still being done between the queues.

In FAIR-FAIR scheduling policy, the fair amount of resources will be shared by the jobs that are running in the same queue.

FAIR-DRF scheduling policy is devised by Ghodsi et al. [7]. In FAIR-DRF scheduling policy, DRF evaluates the resources shared by each user, finds out the maximum of it, and calls it as a dominant resource of the user. The idea is to make uniform resource sharing among the users through equalizing the resources like CPU and Memory.

## 4 Experimental Environment, Workload, Performance Measure, and Queue Configuration

In this paper, we evaluate the performance of two Hadoop schedulers by using three built-in scheduling policies (i.e., FIFO, Fair, and DRF) and test it in context to different queue settings. The performance metrics include six dimensions, which is the data locality, latency, completion time, turnaround time, and the CPU and memory utilization. For our experiment, we implemented Hadoop (2.7.2) cluster. The cluster consists of 1 master node and 11 slave nodes. The important information of the cluster configuration is shown in Table 1.

**Table 1** Hadoop cluster configuration

| Configured parameter | Master node | Cluster nodes |
|---|---|---|
| Nodes (in cluster) | 1 (Namenode) | 11 (Datanode) |
| Network | 1 Gbps | 1 Gbps |
| CPU | Pentium Dual-Core CPU @ 3.06 GHZ * 2 | Pentium Dual-Core CPU @ 3.06 GHZ * 2 |
| Cache | L1—64 KBL2—2 MB | L1—64 KBL2—2 MB |
| RAM | 4 GB | 22 GB (11 Nodes * 2 GB per node) |
| Disk | 500 GB SATA | 500 GB SATA |
| Block size | 128 MB | 128 MB |
| CPU cores | 2 | 22 (11 Nodes) |
| OS | Ubuntu 14.04 LTS | Ubuntu 14.04 LTS |

**Table 2** Total workload
information

| Job | #Maps | #Reducers | Job Size |
|-----|-------|-----------|----------|
| WordCount | 24 | 10 | 2 GB |
| WordMean | 24 | 1 | 2 GB |
| TeraSort-1 | 15 | 10 | 2 GB |
| TeraSort-2 | 25 | 10 | 3 GB |
| Pi-1 | 20 | 1 | 1,00,000 samples |
| Pi-2 | 50 | 1 | 1,00,000 samples |

Six big data jobs are chosen for this experiment. Those are WordCount, Word-Mean, TeraSort, and PiEstimator. Here, WordCount, WordMean, and TeraSort are CPU-intensive jobs, while PiEstimator is more memory-intensive job.

WordCount calculates the total n of words in a file, while WordMean calculates the average length of words in a given file. TeraSort performed on data generated by TeraGen, and PiEstimator estimates the value of Pi. Our workload is combination of CPU and memory-intensive jobs to check the performance of Hadoop schedulers under the experiment environment in very effective way. Details of each job are given in Table 2.

The performance of the Hadoop schedulers has been evaluated by considering the following parameters:

**Locality**: HDFS maintains the copy of the data splits across the datanodes. When any job is executed, MapReduce divides the job among multiple tasks, which is submitted for execution on multiple datanodes. Each mapper requires the copy of data to process, if the data is not available on the same datanode, it will find and bring the copy of data from other datanode over the network, to the node where it is required. So if the data is available on the same node, it is called data local; if it is available on the same rack, it is called rack local; and if both scenarios fail, then it will be copied from the different rack data. So if job finds most of the data locally, then completion time of job would be better compared to the rack local or different rack data (The data found on another node but within same cluster is known as data on same rack. If data is residing on a datanode in a different cluster, it is referred as a different rack).

**Latency**: It is the time that the job has to wait, until getting scheduled, after the job is submitted.

**Completion Time**: It is the difference of finish time and start time of the job. It is the sum of actual execution time of the job and the waiting time, if any.

**Turnaround Time**: It is the total amount of time elapsed between submission of the first job and until the completion of the last job, in the queue.

**CPU and memory utilization**: Hadoop counters provide the time spent by the job on CPU and total memory bytes utilized by the job.

For experiment purpose, three types of job queues have been configured: (a) single queue, (b) multi-queue, and (c) mixed multi-queue.

**Table 3** Workload assignment to the queue

| Multi-queue | | Mixed Multi-queue | |
|---|---|---|---|
| Queue | Jobs | Queue | Jobs |
| A | WordCount | A | WordCount |
| A | WordMean | A | TeraSort-1 |
| B | TeraSort-1 | B | Pi-1 |
| B | TeraSort-2 | B | TeraSort-2 |
| C | Pi-1 | C | WordMean |
| C | Pi-2 | C | Pi-2 |

**Single Queue**

In single queue, all the resources of a cluster will be used by one queue only. All jobs will be entered and will be scheduled according to the scheduling policy and resources available. Capacity scheduler will be configured with FIFO, while fair scheduler can be configured with FIFO, FAIR, or DRF policy. These four schemes (Cap-FIFO, Fair-FIFO, Fair-FAIR, and Fair-DRF) will be evaluated based on discussed six variables using workload listed in Table 2.

**Multi-queue**

Here for our experiment, we have considered three queues where each queue will be running similar kind of application. Three queues named "A", "B", and "C" is configured with 30%, 40%, and 30% resources, respectively, of the total resources available. These queues have been kept soft so that each queue can use 2 times of their configured capacity for elasticity purpose. Jobs are allocated to the queues as per given in Table 3.

**Mixed Multi-queue**

In this case, to test the performance variation, we change the jobs to different queues. If we put different types of application in the same queue, then what will be the effect on to the performance and under that circumstance performance is measured. For our experiment, mixed jobs are allocated to the queues as per Table 3.

## 5   Performance Evaluation

For our experiment, each application has been executed 5 times to validate the results of the evaluated performance. Performance evaluation is carried out without changing its default settings except for the queue settings of the schedulers.

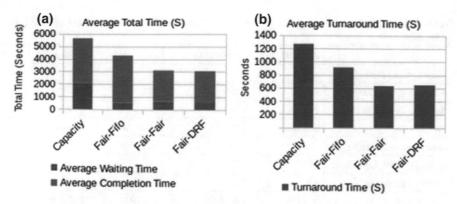

**Fig. 4** **a** Average total time, **b** Average turnaround time

## 5.1 Single Queue

All six big data application entered in a single queue with normally 2 s of delay. By looking at the results in Fig. 4a, we can say in Cap-FIFO scheduler job waiting time and completion time are comparatively very high as compared to other three scheduling policies. Fair-FIFO has less waiting time as job gets scheduled by fair policy but completion time is more compared to Fair-Fair and Fair-DRF policies. While Fair-Fair and Fair-DRF perform better, it has an almost similar job waiting time and completion time. As shown in Fig. 4b in terms of turnaround time, also Fair-Fair and Fair-DRF outperform other scheduling policy.

As shown in Table 4 in terms of data locality, 74.40% tasks are data local in Fair-DRF, which means Fair-DRF finds maximum data task as data local out of the total tasks launched. Moreover, Fair-DRF gives more resource efficiency as total task launched by Fair-DRF is less compared to others. In terms of CPU time and memory usage shown in Fig. 5, Fair-Fair is better as it utilizes less CPU time and effectively uses the physical memory.

## 5.2 Multi-queue

As shown in Fig. 6a, we can say in Cap-FIFO scheduler job waiting time is less but completion time is comparatively very high as compared to other three scheduling policies. Fair-FIFO performs better in terms of the job waiting time and completion time in comparison to Fair-Fair and Fair-DRF policies. As shown in Fig. 6b in terms of turnaround, also Fair-FIFO is more efficient than Fair-Fair and Fair-DRF scheduling policies.

In terms of data locality, as shown in Table 4, 76.33% tasks are data local in Fair-Fair, which means Fair-Fair finds maximum data task as data local out of the total

**Table 4** Data locality (percentage shows how many % of data found data local out of total task launched)

|                          | Capacity | Fair-FIFO | Fair-Fair | Fair-DRF |
|--------------------------|----------|-----------|-----------|----------|
| *Single queue locality*  |          |           |           |          |
| Average Total_Launched   | 201      | 179       | 176       | 168      |
| Average Data_Local       | 126      | 112       | 114       | 125      |
| Percentage               | 62.69%   | 62.57%    | 64.77%    | 74.40%   |
| *Multi-queue locality*   |          |           |           |          |
| Average Total_Launched   | 211      | 178       | 169       | 172      |
| Average Data_Local       | 93       | 121       | 129       | 119      |
| Percentage               | 44.08%   | 67.98%    | 76.33%    | 69.19%   |
| *Mixed Multi-queue locality* |      |           |           |          |
| Average Total_Launched   | 196      | 173       | 169       | 166      |
| Average Data_Local       | 85       | 122       | 122       | 120      |
| Percentage               | 43.37%   | 70.52%    | 72.19%    | 72.29%   |

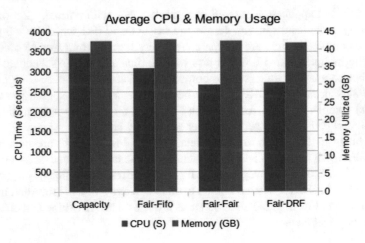

**Fig. 5** CPU and memory usage

tasks launched also Fair-Fair gives more resource efficiency as total task launched by Fair-Fair is less compared to others. In terms of CPU time and memory usage shown in Fig. 7, Fair-FIFO is better as it utilizes less CPU time and effectively uses the physical memory.

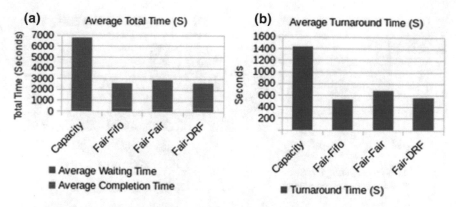

Fig. 6 **a** Average total time, **b** Average turnaround time

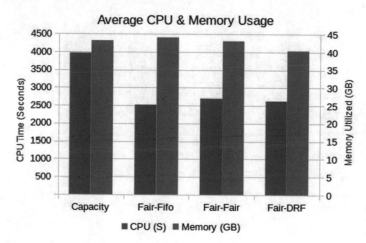

Fig. 7 CPU and memory usage

## 5.3 Mixed Multi-queue

As shown in Fig. 8a, we can say in Cap-FIFO scheduler job waiting time is less but completion time is comparatively very high as compared to other three scheduling policies. Fair-Fair and Fair-DRF perform similar in terms of the job waiting time but if we consider total time, then Fair-DRF performs better than rest of all scheduling policies. As shown in Fig. 8b in terms of turnaround, also Fair-DRF is more efficient than Fair-Fair and Fair-DRF scheduling policies.

As shown in Table 4 in terms of data locality, 72.29% and 72.19% tasks are data local in Fair-DRF and Fair-Fair, respectively, which means Fair-Fair and Fair-DRF find maximum data task as data local out of the total tasks launched also Fair-DRF gives more resource efficiency as total task launched by Fair-DRF is less compared

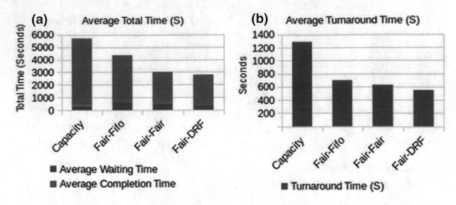

Fig. 8  **a** Average total time, **b** Average turnaround time

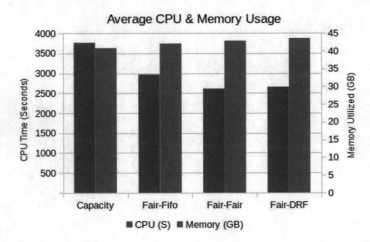

Fig. 9  CPU and memory usage

to others. In terms of CPU time and memory usage shown in Fig. 9, Fair-Fair and Fair-DRF perform similarly.

## 6  Conclusion

In this paper, various Hadoop scheduling algorithms have been evaluated based upon latency time, completion time, and data locality. For experiment purpose, six big data applications have been implemented using three different scheduling queue configurations such as single queue, multi-queue, and mixed multi-queue. Various experiments were conducted by fine-tuning scheduling policy for Hadoop environment. The results of the experiments are presented here which may be useful in selecting a scheduler and scheduling policy depending upon application that one wants to run. Based on the results, the following conclusions can be drawn:

1. In single queue, Fair-DRF outperforms in terms of execution time and effective resource usage capacity as compared to other three. The only factor that it lacks is that CPU usage time is a bit high compared to Fair-Fair scheduler.
2. In multi-queue, Fair-FIFO is the best option if we consider workload waiting time, completion time, turnaround time, and CPU usage. Fair-Fair is better only when resource utilization is important.
3. In mixed multi-queue, Fair-DRF is the most appropriate choice with respect to resource utilization and workload execution performance.

The results which are drawn here can be application dependent and future researcher can test the same with different application types. The future enhancements can include testing the impact of delay scheduler on capacity and fair scheduler. It is presumed that delay scheduling may make a significant impact on the performance of the scheduler.

# References

1. Forbes Welcome. https://www.forbcs.com/sites/gilpress/2014/09/03/12-big-data-definitions-whats-yours/#487d104413ae. Accessed 30 May 2017
2. Hadoop. http://hadoop.apache.org. Accessed 30 May 2017
3. Dean, J., Ghemawat, S.: MapReduce: simplified data processing on large clusters. Commun. ACM **51**(1), 107–113 (2008)
4. Kalavri, V.: Performance optimization techniques and tools for data-intensive computation platforms. www.diva-portal.org/smash/get/diva2:717735/fulltext01.pdf. Accessed 1 May 2017
5. Hadoop Capacity Scheduler. https://hadoop.apache.org/docs/r2.7.2/hadoop-yarn/hadoop-yarn-site/CapacityScheduler.html. Accessed 20 May 2017
6. Hadoop Fair Scheduler. https://hadoop.apache.org/docs/r2.7.2/hadoop-yarn/hadoop-yarn-site/FairScheduler.html. Accessed 20 May 2017
7. Ghodsi, A., Zaharia, M., Hindman, B., Konwinski, A., Shenker, S., Stoica, I.: Dominant resource fairness: fair allocation of multiple resource types. NSDI **11**, 24–24 (2011)

# Energy-Aware Prediction-Based Load Balancing Approach with VM Migration for the Cloud Environment

Durga Patel, Rajeev Kumar Gupta and R. K. Pateriya

## 1 Introduction

In the IT field, a new technology is emerging regularly and adoption of these technology demands several investments on infrastructure that is no't possible for all the enterprises. Cloud computing has resolved these issues by providing services on-demand basis that may be accessed through web on pay-per-use basis. The cloud is one of the most demanding technologies in IT industries [1, 2]. The cloud has created a new look to align IT and business visions. It is providing Software-as-Service (Saas), Platform-as-Service (Paas) and Infrastructure-as-Service (Iaas) in a very virtualized cloud atmosphere. The cloud computing power is created through distributed computing and the advanced communication networks. Cloud works on the principle of virtualization of resources with on-demand and pay-as-you go model policy. It is originated from the large-scale Grid atmosphere. It is a technology of computing that is extensively utilized in today's business moreover as a society. Applications run within the cloud, the user will access it anywhere at any time through an associate Internet-enabled mobile device or a connected PC.

Due to some attractive features, demand for the cloud resources are increasing gradually. In order to fulfill this demand huge number of servers are used which runs several VMs. These servers consume large amount of energy for their operational and cooling function, which leads to emitting large amount of $CO_2$. This energy consumption can be minimized by reducing the number of running server in the data center [3]. The key thought of the cloud computing is Virtualization. Virtualization

D. Patel (✉) · R. K. Gupta · R. K. Pateriya
Department of CSE, MANIT, Bhopal, India
e-mail: durgapatel28@gmail.com

R. K. Gupta
e-mail: rajeevmanit12276@gmail.com

R. K. Pateriya
e-mail: pateriyark@gmail.com

© Springer Nature Singapore Pte Ltd. 2019
R. K. Shukla et al. (eds.), *Data, Engineering and Applications*,
https://doi.org/10.1007/978-981-13-6351-1_6

has become a very popular valuable and standard technology for cloud computing environment. Virtualization is the concept that abstracts the coupling between the hardware and the operating system and enhances the computing power without installing additional hardware infrastructure. Virtual Machine Monitor (VMM) is a great tool for administrator which is responsible for the separation between software and hardware and add the flexibility to migrate the VM from one PM to another. Virtual Machine Migration allows load balancing which leads to increase the resource utilization and energy saving [4]. VM migrations techniques are classified mainly into three categories:

  i.   Energy-Efficient Migration Techniques
 ii.   Fault-Tolerant Migration Techniques
iii.   Load Balancing Migration Techniques.

**i. Energy-Efficient Migration Techniques**
The energy consumption of data center is mostly estimated by the utilization of the servers and their systems. The servers usually went up to 70% of their highest power consumption, even at their low utilization level. Hence, there is a necessity for migration methods that preserves the energy of servers by ideal resource utilization [5].

**ii. Load Balancing Migration Techniques**
The main goal of load balancing migration technique is to distribute the load across the physical server to enhance the scalability of servers in a cloud environment. Load balancing helps in evading bottlenecks, minimizing resource consumption, enhancing scalability, implementation of fail-over, over-provisioning of resources, etc. [4].

**iii. Fault-Tolerant Migration Techniques**
Fault tolerance gives the authority to the virtual machines to do its job continuously if any part of system failure occurs. In this technique, virtual machine migrates from one physical server to another physical server and the migration is done based upon the prediction of failure. The advantages of the fault tolerance migration technique are to avoid performance degradation of applications, improve the availability of the physical server [5, 6].

This paper presents an energy-aware VM migration-based load balancing approach for the cloud, which will reduce the energy consumption and minimizes number of VM migration while keeping efficient utilization of resources. In order to minimize the number of migration due to the temporary peak load forecasting approach is used.

## 2   Literature Survey

Shivani Gupta et al. (2015) proposed a load balancing approach based on Dynamic Threshold [7] in cloud computing. Depending on the systems, current state load balancing problem can be categorised into two approaches namely: Static approach and Dynamic approach. In the static approach, value of the lower and upper thresholds are fixed based on the server configuration whereas in dynamic threshold approach value of the threshold changes with time. Static threshold is more suitable for the cloud due to dynamic nature of the cloud services.

Raj Kumar Buyya et al. (2010) proposed Modified Best-Fit Decreasing (MBFD) algorithm for the efficient utilization of the cloud resource [3]. This paper introduced server consolidation approach for reducing the energy consumption. Fixed lower and upper thresholds are used to find the status of the PM. If the status of the PM is underloaded, all VMs are shifted or migrated to the other PM whereas if the status of the PM is overloaded then of the Best-Fit Decreasing (BFD) algorithm finds the VM where power difference between after and before allocation of VM is minimum. This approach seems good not mitigate the migration due to the temporary peak load.

Akshay Jain et al. (2013) proposed a load balancing approach, which is based on the threshold band [8]. Instead of using single upper threshold, this approach uses the threshold band to define the overloaded status of the PM. The decision of VM selection for the VM migration is depending on two factors: threshold value and mean deviation of all hosts. The value of the threshold band is different for each host depending on the type of application running on it. This approach had not used any prediction approach, which leads to increase in the migration due to the temporary peak load. Iniya Nehru et al. (2013) proposed prediction-based load balancing approach for cloud. In this technique, Neural Network based prediction approach is used to predict the load on PM, which uses past historical load data of each PM [9]. Nada M. Al Sallami et al. (2013) proposed technique for load prediction based on the artificial neural network. For Workload distribution across the nodes back-propagation learning algorithm is used and trained, feed forward Artificial Neural Network(ANN). This proposed technique is very efficient and simple when effective training or learning data sets are used. Artificial Neural Network uses prediction policy and in advance it predicts the demand of resources than start allocating the resources according to that demand [9].

S.K. Mandal et al. (2013) proposed an energy-efficient Virtual Machine placement algorithm for on-demand access to infrastructure resource in cloud computing which reduces the VM allocation time and optimize the resource utilization. In order to improve the resource utilization instead of directly sending the VM requests to the host machine, this algorithm creates the binary search tree of requested VM and then send it to the VM scheduler. To implement the best-fit strategy, the PM's are also organized in binary search tree. Now VM scheduler selects the maximum requirement VM from the VM tree and searches for the host that would be best fit in requirement. By using the binary search tree data structure, the host searching time will take average $O(\log N)$ where N is the number of physical machines [10]. Main limitation of this approach is that it is a single-dimensional best-fit approach, which is suitable for the single-dimensional cloud computing environment, but in cloud computing user can requests for multiple resources simultaneously. Moreover, in this algorithm hosts are organized in binary search tree if binary search tree is left skewed or right skewed then it will take O(n) search time in worst case to find the best-fit host because best-fit host will be in the leaf node, i.e., the last node of the binary search tree [10].

## 3 Proposed Methodology

Cloud is the commercial model where proper utilization of physical resources is the prime requirement of the provider. This paper proposed a load balancing approach, which optimizes the cloud services by increasing the resource utilization. The main components of cloud are data center, server (PM), and VM where each data center can have multiple PM and each PM can run a number of VMs. When user demands for the resources hypervisor (VMM) creates a VM and assigns to the user. Since in cloud, resource requirement of the VM change dynamically so it may lead to a situation where PM is unbalanced. In order to balance these PM and increase the resource utilization an efficient load balancing approach is required that takes the decision according to the situation.

This paper proposed double threshold based dynamic load balancing approach, where thresholds are calculated based on the host utilization. VMs are migrated from one PM to another based on these threshold values. Since the number of migrations affect the performance of cloud services so to minimize the number of migrations is also one of the prime requirement of the cloud provider. In order to achieve this proposed approach, we use the prediction model. Table 1 shows the symbol used in the proposed approach.

Following steps are involved in the proposed load balancing approach:

  i. Calculate load on the PM and VM.
 ii. Calculate the upper and lower threshold to find the overloaded and underloaded condition.
iii. Use prediction model to minimize the number of migrations
 iv. Select suitable VM for the migration by using optimization technique.

**Table 1** Description of the various symbols

| | |
|---|---|
| $PM^{load}$ | Load on the host or server |
| $VM^{cpu}$ | CPU utilization of the VM |
| $VM^{bw}$ | Bandwidth utilization of the VM |
| $VM^{ram}$ | RAM utilization of the VM |
| $VM^{load}$ | Load on the VM |
| n | Number of VMs running in single PM |
| M | Number of PM in the data center |
| $PM_{upper}$ | Upper threshold of the PM |
| $PM_{lower}$ | Lower threshold of the PM |

v.   Select suitable host to place the selected VM.

## 3.1   Load Calculation for PM and VM

One of the challenging tasks in the cloud is to calculate the load on physical and virtual machine because of its dynamic nature. Load on the PM mainly depends on the three parameters namely CPU, memory, and bandwidth. CPU utilization, RAM utilization and Bandwidth utilization can be calculated by using the following formulae.

$$VM^{cpu} = \frac{\text{Total Requested Mips}}{\text{Total Mips of PM}}$$

$$VM^{bw} = \frac{\text{Bandwidth used by VM}}{\text{Total Bandwidth of host}}$$

$$VM^{ram} = \frac{\text{Ram used by VM}}{\text{Total RAM of host}}$$

Since each PM can have a number of VMs and each VM running some applications, so total load on the PM is the summation of load of all VM running in the server. This paper considers only CPU utilization of the VM for calculating load on the PM. Hence,

$$VM_{load} = VM^{cpu}$$

Total load on the host is given by

$$PM^{load} = \frac{\sum_{i=1}^{n} VM_i^{cpu}}{\text{Total MIPS of the PM}}$$

Where n is number of running VM in the $i$th PM.

## 3.2   Lower and Upper Threshold

The performance of the PM depends on its load, which decreases with increase in load. So to define the amount of resources that can be used lower and upper thresholds are used. These thresholds define the overloaded and under loaded situation. A PM is underloaded if PM load is lower than its lower threshold and if the load is more than upper threshold than PM is called overloaded. Thresholds can be Static as well as dynamic. In case of static threshold, lower and upper thresholds are fixed thus, these value does not change with time, while in the dynamic threshold, value of the lower and upper thresholds keep changing (usually lower threshold value is kept constant because it is pre-decided according to system configuration. Due to this reason, the proposed approach uses fixed value of lower threshold, i.e., 20. Appropriate value of upper threshold plays a vital role in the load balancing and highly influence the number of migration.

The lower value of upper threshold will increase the resource wastage whereas higher value will lead to higher number of migrations. Hence, to balance the tradeoff between resource wastage and number of migration is a challenging task. In order to find upper threshold, the proposed approach takes the average of all PM load available in the data center. The main component of the PM is the CPU, RAM and bandwidth, but the previous studies say that CPU is the core element of the cloud resources and performance of the cloud services mainly depends on the CPU. Due to this, proposed approach considers only CPU load for evaluating the performance. In this approach, upper load on the PM is propositional to the average load of the data center. To ensure the high performance, upper threshold should not be less than 70% and not exceed 95% of the total capacity of the PM.

$$PM_{upper} = \frac{\sum_{i=1}^{m} PM^{load}}{m}, \text{Hence}$$

$$PM_{lower} = 0.2$$

$$PM_{upper} = 1 - x * T_{avg}$$

$$PM_{upper} = 1 - \frac{5 * T_{avg}}{100}$$

$$70 < PM_{upper} < 90$$

### 3.3   Load Prediction Using Artificial Neural Network

Due to the dynamic behavior of the cloud environment, if the VM is migrated on the basic of lower and upper threshold then it will lead to the unnecessary migration due to the temporary peak load. This migration can be avoided by using some forecasting method. Our proposed approach uses a prediction approach, which is based on the ANN for prediction the future load on the PM. In ANN backpropagation algorithm is used with Multi-layer perceptron model for finding the predicted load. In neural network, a basic model of three layers with input nodes are used for implementation of ANN. The load prediction of each PM is done by training the perceptrons, for training these perceptrons backpropagation algorithm is used. In order to make effective prediction of load the ANN is trained with a data set. The training is finished by implementing the set input and target data (Fig. 1).

After training the network, load of each host is given as input to ANN, which gives the predicted load in the output layers. ANN is divided into three layers first, the input layer, second, the hidden layer, third is the output layer. Each layer has their own calculations and in combination produces an optimized result which is near to

**Fig. 1** ANN architecture

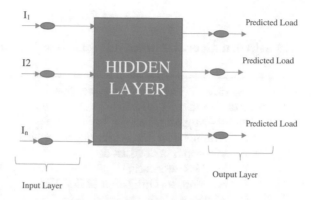

the real result. The input layer takes the CPU utilization value in terms of MIPS and number of VM available at the host. These two values are given and we make an activation function in order to activate the ANN for functioning.

$$\text{input layer: } i = g\left(\sum_{i=0}^{n} x_i w_i + b\right)$$

Hidden layer of ANN implements an exponential function in order to get trained and circulates the data from input layer to hidden layer and vice versa.

Output layer contains the neuron, which produces the output using following formula:

$$\text{output layer} = \frac{1}{1 + e^{-i}}$$

The output layers give the predicted load, which is used to trigger the VM migration polices.

### 3.4 VM Selection Algorithm

When the status of the PM is overloaded or underloaded some/all VM need to be migrated to balance the PM. Since each host runs a number of VM and these VM change their resource requirement dynamically, so VM selection is the critical job in the cloud. A wrong VM selection may lead to increase the total migration time and downtime. When the status of PM is underloaded all VM running on that PM has to be migrated to save energy whereas if the status of PM is overloaded proposed approach migrate the VM till the PM is balanced. In the overloaded case, we migrate the VM whose size is greater than or equal to the difference between upper threshold and host utilization. VM with higher $VM_{size}$ will be chosen first.

$$VM_{size} = PM_{upper} - PM^{load}$$

**VM Selection Algorithm when the PM is Overloaded**

Algorithm 1 VM Selection Algorithm
Input - VmList,HostList
Output - MigrationList
1.  Arrange host into decreasing order according to utilization
2.  for each h in hostList do
3.  hostUt ←host.util()
4.  bestVmUt←Utilization of first VM
5.  whilehostUt>host.upThresh do
6.  for each vm in vmList do
7.  diff ← hostUt − host.upThresh
8.  if vm.util() > diff then
9.  $T_{avg}$ ← vm.util()
10. if $T_{avg}$<bestVmUt then
11. bestVmUt ← $T_{avg}$
12. bestVm ←vm
13. else
14. if bestVmUt= First VM  then
15. bestVm ←  vm
16. break
17. hostUt ← hostUt − bestVm.util()
18. migrationList.add(bestVm)
19. vmList.remove(vm)

**VM Selection Algorithm when the PM is Underloaded (VM Consolidation)**

1.  if hotUtil < lowThresh() then
2.  Migrate all VM from the host

## 3.5  Select the Host to Place the Selected VM

Final PM Selection is the most critical step in the VM migration, because it affects
the overall performance of the system. Wrong selection of the PM may increase the
number of VM migration as well as resource wastage. A VM needs to be placed in
two different situations either for the new VM or migrated VM. In both cases, the
proposed approach chooses VM in such a way it will minimize energy consumption
and resource wastage. In the proposed algorithm, first we find the overloaded host
and underloaded host by the use of upper threshold and lower threshold after that

selection of VM and migration is done after this load balancing by using DDT algorithm and then data sent to the ANN, it gives predicted host CPU utilization list and this CPU utilization list is used for calculating upper threshold and finding the overloaded host. So, in second iteration VM migration is done according to the predicted load given by ANN and this process is continued.

## PM Selection Algorithm to Place New VM

Algorithm 3 HostSelection for VM
Input - HostList, vmList
Output -  Allocation of VMs
1.    Sort all PM according their utilization
2.    for eachvm in vmList do
3.    for each host in hostList do
4.    if Host_Load<=H_UTD &&Host_Load>=H_LTD
5.    Assign VM to the host where less increment in the power
6.    else
7.    Activate new host and assign VM to that host

## Energy-Aware Optimized PM Selection Algorithm to Place Migrated VM

**Input**: Host CPU Utilization List., **Output**: Minimizing VM Migration
1.    Host_Utl_List in MIPS
2.    if (Host_Utlgreater thanH_LT &&Host_Utl  less than H_UT) then
3.      Wait for time t sec
4.    elseif(Host_Utl  > H_UT) then
5.    Select VM from VM Migration List-1 &migrate VM with Host_Utl between UT and LT and
      (UtliAfterAlloc greater than PrvUtli this Host ) then
6.        Migration successful
7.        if
8.            successful then
9.              wait for time t
10.      else
11.         repeat
12.  elseif (Host_Utl < H_LT) then
13.      Select VM from VM Migration List-2 & migrate VM with Host_Utl > LT and
          (UtliAfterAlloc > PrvUtli this Host ) then
14.      Migration successful
15.      if
16.        successful then
17.  send the data to ANN
18.      else
19.        repeat
20.    ANN gives predicted Host CPU Utilization List
21.    repeat
22.  form step-1

# 4 Performance Evaluation

## 4.1 Experiment Setup

For performing experiments physical machine has a 2.40 GHz Intel(R) core(TM) i5 processor, 8 GB RAM, 1 TB hard disk, and 64-bit Operating System Windows 10 Professional. In our experiment latest version of CloudSim toolkit, CloudSim 3.03 [10–12] is used to for simulation of Cloud computing scenarios and MATLAB 2015a is used for implementing ANN.

## 4.2 Experimental Results

We have analyzed different scenarios by taking different numbers of hosts, different numbers of virtual machines and various number of cloudlets (tasks), i.e., load to evaluate the performance of proposed algorithm. It is essential to use workload traces from real system. In this simulation, we have taken different work load. We have plotted different graphs based on different workloads between two strategies, proposed method DDT with ANN and the existing method DDT (Dynamic Double Threshold). The performance and accuracy of the proposed approach is calculated for two parameters (i.e., Energy Consumption(Energy_Con), Number of VM migrations(VM_Mig)). These results can vary according to the different time slots (Table 2).

Figure 2 represents a comparison between VM migration and energy consumption for dynamic double threshold and double threshold with ANN. It is evident from the graph that as the number of VM migration increases energy consumption increases

**Table 2** Performance of the various approaches for different number of PM

| No. of hosts | No. of VMs | No. of cloudlets | VM_Mig (DDT) | VM_Mig (ANN) | Energy_Con (DDT) | Energy_Con (ANN) |
|---|---|---|---|---|---|---|
| 5 | 10 | 20 | 3 | 3 | 0.12 | 0.12 |
| 10 | 20 | 30 | 16 | 16 | 0.26 | 0.26 |
| 15 | 30 | 40 | 28 | 26 | 0.34 | 0.30 |
| 20 | 40 | 50 | 33 | 31 | 0.48 | 0.40 |
| 25 | 50 | 60 | 54 | 50 | 0.62 | 0.52 |
| 30 | 60 | 70 | 57 | 50 | 0.77 | 0.54 |
| 35 | 70 | 80 | 66 | 62 | 0.90 | 0.81 |
| 40 | 80 | 90 | 77 | 77 | 0.98 | 1.00 |
| 45 | 90 | 100 | 101 | 81 | 1.12 | 0.01 |
| 50 | 100 | 110 | 107 | 85 | 1.23 | 1.20 |

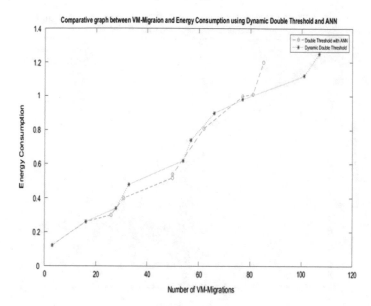

**Fig. 2** Comparative graph between VM migration and energy consumption using DDT and DDT with ANN

in both of the load balancing algorithms. The graph between VM migration versus energy consumption is plotted by continuously increasing the number of hosts, VM and cloudlets. Analysis shows that number of VM migration is less for double threshold with ANN, consequently energy consumption will also be less.

Figure 3 represents number of VM migrations and energy consumption with the condition of increasing number of cloudlets at constant number of hosts and VMs. In this case, number of VM migration and energy consumption for dynamic double threshold is greater than double threshold with ANN. There is one outlier with the trend, with number of migration equivalent to 16. This condition arises due to external factors existing in cloud (Table 3).

Figure 4 represents comparison between number of VM migration and energy consumption for dynamic double threshold and double threshold with ANN, keeping number of hosts and cloudlets constant. Both algorithms perform equivalently in this situation (Table 4).

Figure 5 represents a comparison between VM migration and energy consumption for dynamic double threshold and double threshold with ANN. The above case is considered for different time intervals keeping number of hosts, VMs and cloudlets constant. Dynamic double threshold's performance is poorer than double threshold with ANN.

The workspace area shows in Fig. 6 the variables and parameters related to ANN implementation. The values of different parameters are also known to us. From here, we can plot graphs for different parameters available like value of input and output, Root Mean Square Error rate, etc. (Fig. 7).

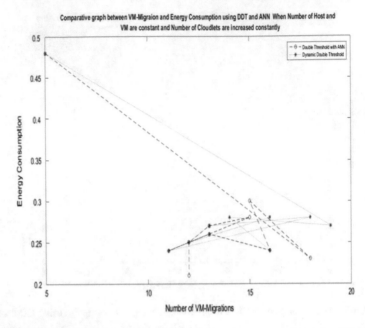

**Fig. 3** Comparative graph between VM migration and energy consumption using DDT and ANN (Number of host, VMs are constant and cloudlets are increased constantly)

**Table 3** Performance of the various approaches for 10 number of PM

| No. of hosts | No. of VMs | No. of cloudlets | VM_Mig (DDT) | VM_Mig (ANN) | Energy_C (DDT) | Energy_C (ANN) |
|---|---|---|---|---|---|---|
| 10 | 5 | 50 | 0 | 0 | 0.06 | 0.06 |
| 10 | 10 | 50 | 8 | 8 | 0.13 | 0.13 |
| 10 | 15 | 50 | 12 | 12 | 0.18 | 0.18 |
| 10 | 20 | 50 | 13 | 13 | 0.25 | 0.25 |
| 10 | 25 | 50 | 21 | 21 | 0.36 | 0.36 |
| 10 | 30 | 50 | 20 | 20 | 0.42 | 0.42 |
| 10 | 35 | 50 | 28 | 25 | 0.45 | 0.43 |
| 10 | 40 | 50 | 30 | 26 | 0.51 | 0.47 |
| 10 | 45 | 50 | 28 | 28 | 0.46 | 0.46 |
| 10 | 50 | 50 | 30 | 29 | 0.48 | 0.47 |

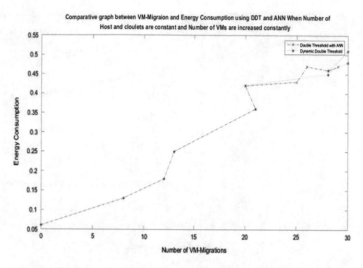

**Fig. 4** Comparative graph between VM migration and energy consumption using DDT and DDT with ANN (Number of host, cloudlets are constant and VMs are increased constantly)

**Table 4** Performance of the various approaches for 20 number of PM

| No. of hosts | No. of VMs | No. of cloudlets | VM_Mig (DDT) | VM_Mig (ANN) | Energy_Con (DDT) | Energy_Con (ANN) |
|---|---|---|---|---|---|---|
| 20 | 35 | 35 | 32 | 27 | 044 | 0.41 |
| 20 | 35 | 35 | 27 | 25 | 0.45 | 0.45 |
| 20 | 55 | 35 | 27 | 26 | 0.41 | 0.40 |
| 20 | 35 | 35 | 29 | 27 | 0.45 | 0.41 |
| 20 | 35 | 35 | 28 | 29 | 0.40 | 0.41 |
| 20 | 35 | 35 | 22 | 24 | 0.45 | 0.40 |
| 20 | 35 | 35 | 33 | 28 | 0.43 | 0.41 |
| 20 | 35 | 35 | 28 | 27 | 0.41 | 0.41 |
| 20 | 35 | 35 | 33 | 26 | 0.41 | 0.40 |
| 20 | 35 | 35 | 30 | 28 | 0.42 | 0.41 |

By using ANN, the percentage of accuracy is "94.5555%" which is better prediction and this accuracy level will increase more and more to nearly about 100%, i.e., 99.9999%. This much of accuracy will be achieved as many input values are passed through ANN.

The result shows that proposed method Energy-Aware VM Migration-based Load Balancing Approach for the Cloud Computing give the result better than existing method in all scenarios. It gives less number of migrations as compared to existing method in all conditions. It is achieved by future load prediction by using ANN. ANN shows 94.5555% accuracy of load prediction.

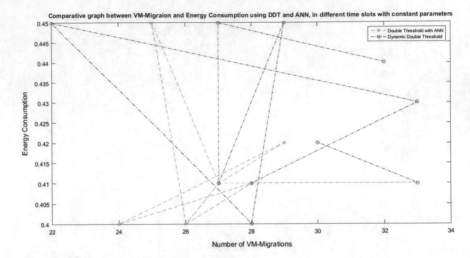

**Fig. 5** Comparative graph between VM migration and energy consumption using DDT and DDT with ANN in different time slots with constant parameter

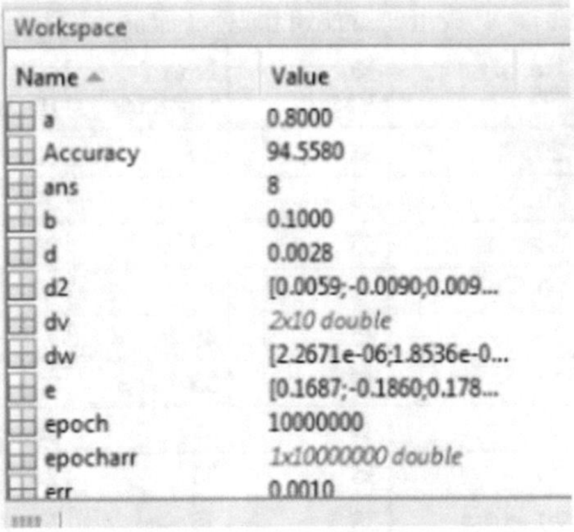

**Fig. 6** Parameters related to ANN

**Fig. 7** Accuracy of ANN

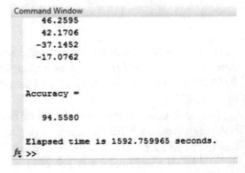

## 5 Conclusion and Future Work

In this research work, we proposed an Energy-Aware VM Migration-based Load Balancing Approach for the Cloud Computing. Our work implements ANN with Dynamic Double Threshold algorithm for calculating VM migration and energy consumption with CPU utilization as a parameter. It not only minimizes number of VM migrations but also reduces energy consumption. The comparative graphs between VM migration and energy consumption explains that, if we integrate ANN with existing load balancing algorithms then the number of VM migrations and energy consumption reduces marginally but it saves lots of power consumption of cloud service providers, as the time pass on. The RMS value for calculating the output by ANN shows maximum accuracy rate.

In the future, we can extend our implementation from CPU utilization as a parameter to CPU, bandwidth, RAM altogether as a parameter. This will improve efficiency of load calculation which minimizes VM migration and energy consumption. Since, ANN is a learning machine and gives optimized result after each iteration, we can integrate ANN with Cloud server for continuous monitoring of load during different times slots. Consequently, by using real cloud data into ANN, we can improve the efficiency of load balancing approach in future. This approach can be integrated with other existing load balancing algorithms for the best result which any algorithm can get.

## References

1. Buyya, R., et al.: Cloud computing and emerging IT platforms: vision, hype, and reality for delivering computing as the 5th utility. In: Future Generation Computer Systems, pp 599–616. Elsevier Science (2009)
2. Qaisar, E.J., et al.: Introduction to cloud computing for developers key concepts, the players and their offerings. In: Proceeding of IEEE Conference on Information Technology Professional, pp. 1–6 (2012)
3. Beloglazov, A., et al.: Energy-aware resource allocation heuristics for efficient management of data centers for cloud computing. Elsevier J. Future Gener. Comput. Syst. 755–768 (2012)
4. Kapil, D., et al.: Live virtual machine migration techniques: survey and research challenges. In: IEEE 3rd International Advance Computing Conference (IACC), pp. 963–969 (2013)
5. Maziku, H., Shetty, S.: Network aware VM migration in cloud data centers. In: Third GENI Research and Educational Experiment Workshop, pp. 25–28. IEEE (2014)
6. Cerroni, W., Esposito, F.: Optimizing live migration of multiple virtual machines. IEEE Trans. Cloud Comput. (2016)
7. Gupta, Shivani, Tiwari, Damodar: Energy efficient dynamic threshold based load balancing technique in cloud computing environment. Int. J. Comput. Sci. Inf. Technol. 6(2), 1023–1026 (2015)
8. Jain, A., et al.: A threshold band based model for automatic load balancing in cloud environment. In: IEEE International Conference on Cloud Computing in Emerging Markets (CCEM), pp. 1–7 (2013)
9. Nehru, E.I., Venkatalakshmi, B.: Neural load prediction technique for power optimization in cloud management system. In: Proceedings of IEEE Conference on Information and Communication Technologies (2013), 541–544 pp.

10. Mandal, S.K., Khilar, P.M.: Efficient virtual machine placement for on-demand access to infrastructure resource in cloud computing. Int. J. Comput. Appl. 25–32 (2013)
11. Calheiros, R., Ranjan, R., De Rose, C.A.F., Buyya, R.: CloudSim: A Novel Framework for Modeling and Simulation of Cloud Computing Infrastructures and Services, pp. 23–50 (2011)
12. Wickremasingh, B., et al.: A CloudSim-based visual modeller for analyzing cloud computing environments and applications. In: Proceedings 24th International Conference on Advanced Information Networking and Applications (AI NA) (2010)

# Part II
# Network and Securities

# Authentication Process Using Secure Sum for a New Node in Mobile Ad Hoc Network

Praveen Gupta and Pratosh Bansal

## 1 Introduction

Wireless MANETs are deployed by using many independent nodes. These nodes are self-organized nodes. The network has nodes with high mobility. Nodes communicate either directly (when they are in the communication range) or indirectly (using intermediate nodes for data forwarding). MANET can be created at the time of requirement. These networks can easily be established in such areas or situations, where deployment of a wired network is not feasible. With these advantages, a MANET has several disadvantages like node transfers data over an open channel, highly mobile nature of node requires up-to-date location information, nodes are having less computing resources like processing power, less battery power, lesser bandwidth, and may have short life span [1–4].

These features make a MANET more vulnerable. A malicious node tries to access resources without permission. It is the responsibility of the network establisher to design a network with security features. Some MANET does not need high-level security while some may need high security [5]. It depends on the situation for which a MANET is deployed.

Rest of the paper is covered in the following sections. Section 2 describes the security issues in mobile ad hoc network. Section 3 is a literature review. Section 4 explains the proposed algorithm, result, and analysis of results. Section 5 concludes the work.

P. Gupta (✉) · P. Bansal
Information Technology, IET, Devi Ahilya Vishwavidyalaya, Indore, MP, India
e-mail: praveen.gupta@live.com

P. Bansal
e-mail: pratosh@hotmail.com

© Springer Nature Singapore Pte Ltd. 2019
R. K. Shukla et al. (eds.), *Data, Engineering and Applications*,
https://doi.org/10.1007/978-981-13-6351-1_7

## 2 Security Issues in Mobile Ad Hoc Network

Various attacks are possible in MANET. Attacks may occur either from inside or from outside of the network. An intruder is either part of the system or a stranger. It compromises legitimate nodes to get the information [6]. The format of their attack is to access information or data actively or passively. In passive format, an attacker only listens to floating information. In active format, an attacker steals information and either deletes or modifies it. In all the cases, tampering or loss of any data is crucial to complete the goal for which MANET is set up [7]. Sometimes these attacks are done to choke the network activities so that network can be unavailable to its users. This can be done by sending many requests to a genuine node and keep it busy to reject other service requests. It is termed as Denial of Service (DoS) attack [4, 8]. A malicious node does packet flooding in the network and all other nodes waste their resources in the processing of these packets. Many such DoS attacks like jamming, packet injection, blackhole, wormhole, etc., are possible in the network. In some other attacks, malicious node targets routing information. They misguide other nodes in the network by providing false, old or modified route details.

All security features: privacy, authentication, integrity, non-repudiation, and availability cannot be achieved at once. Each system has its pros and cons. Various techniques have been proposed by researchers for the security breaches. Security measures can be applied at all levels of the network. Security measures can be applied either as preventive measures or as reactive measures. It is better to have preventive security measures in the network to avoid the attacker's initial attempt to get the entry or to access the resources in an unauthorized manner.

This paper is focused on one of the security feature authentications to prevent unauthorized entry of node in network and access to resources. Though a legitimate user in the network may get compromised later and perform malicious activities. Authentication in a network assures entry of genuine node. Various ways are used to allow a node in the network. A node can show its credentials and gets an entry in the network. It is easily possible in wired and planned wireless network. For an unplanned mobile ad hoc network, authentication mechanism should be devised in such a manner that it does not put a load on network activities and does not require additional devices.

## 3 Literature Survey

Security issue in a MANET is a critical issue. The major question is how and where to apply security measures to create a fully protected network. Researchers have contributed their work on various types of attack and their countermeasures. In the newborn stage of MANET, researchers were paying attention to general problems like resources availability, channel limitation, and routing issues. Threats were classified in two classes, viz., attacks and misbehavior. Routing attacks like modification, impersonation, fabrication are possible in the network. Authentication, neighbor detection, and its monitoring are some major solutions [1, 2]. Attacks in the MANET were classified as external and internal, attacks at protocol stack layer, cryptographic primitive attacks. Encryption and authentication are the preventive mechanisms and the intrusion detection system is a reactive mechanism [3, 5].

MANETs are used for such situation where quick set up of a wired network is not possible. The set up of MANET also needs security protocols accordingly [7]. Wireless MANET can be deployed using various types of devices. It can be designed in various forms like mesh networks, sensor network, vehicular network, etc. All these networks share common properties of MANET and face common security problems [4, 6].

Secure communication in any network is a challenging job. Authentication is one of the preventive mechanisms that can be achieved by either using encryption or Public Key Infrastructure (PKI), such things need additional set up of special nodes. Authentication can be done in the limited area using key exchange program. It uses the pre-authentication concept to prove the identity of each other in a small area located network [9].

MANET can be developed for its communication between available nodes or it can be developed with a network that connected with the Internet. For this, it requires two levels of authentications, level 1 for the network working without the internet. The node present at one hop count distant of each other and level 2 authentication for a network with the Internet is used when it has a centralized server [10].

The same level of security cannot be achieved in a MANET as available in the wired network. Environmental situations, where network deployed, also give its impact on network working. It raises the question of Quality of Services (QoS) [5, 11]. Major two security factors, authentication, and confidentiality provide high security to a MANET. These factors may be affected by limited resources, node mobility, and no central control nature of MANET. Some techniques for authentication are symmetric cryptography, asymmetric cryptography, threshold cryptography that requires huge resources for computation. Similarly, confidentiality can be attained by using symmetric, asymmetric encryption, and stream cipher [12].

The methods suggested by researchers need additional infrastructure, a centralized server in case of PKI, and extra resources to run heavy encryption algorithms for authentication. This paper concentrates on developing an algorithm that does not need extra infrastructure or resources for short-span unplanned MANETs.

## 4 Proposed Algorithm

Generally, a MANET is always set up in a situation where users do not have enough time to deploy the wired network. The main implementation area of a MANET is an emergency situation. The life span of such a network is very less. It is crucial for an administrator to set up the network with the full plan instead it is important to send readily available nodes to start the work.

Network security is the major concern during its deployment. A MANET is vulnerable to security threats. Threats are ranging from various types of attacks like unauthorized access to network resources, data modification, data deletion, etc. It is required here to secure the flow of information in the network.

A mischievous node tries to obtain floating information either for itself or to deliver it to other nodes. Initially, it is almost impossible to say which node is legitimate or unlawful. There are various ways to save the network from threats. The best preventive mechanism is to restrict such node at the entry level. Authentication of a new node at the entry point of network assures that only legitimate node will work in the network.

The current work proposes an algorithm to authenticate a new node. The algorithm is named as Algo_Authenticate_Node. The main assumptions for this algorithm are: It is designed for small life span of the MANET. It is devised for an unplanned MANET that is deployed for rescue operations like the flood, fire, earthquake, etc. Small life span networks also include conference or meeting. The MANET can start with the minimum of three "key" nodes. These key nodes are used for the authentication process. Other nodes can also be deployed for other tasks.

## 4.1  Algorithm details

This algorithm uses the minimum of three key nodes for authentication. Each node is assigned a large number. Any key node does not have knowledge of what number is assigned to the other two nodes. A secured "sum" (it is the sum of three large numbers) is given to every key node.

When a new node wants to join the network, it must contact the nearest key node. The key node will provide a number to new node, which is the sum of its own large number and a random number "r" (only known to key node). The new node will go to the next available key node and give this sum to it. Next key node will add its own large number to sum and give the new sum to the new node. The same process is done by the third key node.

New node gives the total sum to again the first key node. The first key node will minus the random no "r" from the total sum. If the remaining value is equal to "sum" then the new node has cleared its authentication process and it can be allowed on probation to participate in networks initial activities.

*Algorithm Description*:

*Algo_Authenticate_Node*

---

This algorithm authenticates the request of a node $N_a$ by method secure sum. Assuming key nodes $K_i$ are at least three.

---

*Algo_Authenticate_Node*

Step 1: Start;

Step 2: go to key node 1;

Step 3: r=random ( ); //generates a random big number

        sum = r;

Step 4: for i = 1 to n;

        {

        sum = sum + $k_{ir}$

        }

Step 5: go to node 1;

Step 6: sum = sum-r;

Step 7: if sum != $\sum_{i=1}^{n} k_i$

        then

        discard the node

        otherwise

        insert the node

Step 8: Stop;

---

*Advantages of the algorithm*: first, it does not require many nodes for authentication. The network must have minimum 3 key nodes for authentication. (If two nodes can be used they may easily guess the opponent number). Even, number of nodes will increase the authenticity in the network; second, it is lightweight algorithm and does not require heavy resources and time to compute the process. Third, it is used for a network of a short-span time period. Fourth, it is difficult to guess the individual number as well as the secured sum. Fifth, the algorithm takes three large numbers, which is not possible to break in a short span of time.

The above algorithm has its limitations. First, it is used for short lifetime of the network. Second, if anyone of the key node dies, authentication activity stops. Third, for high mobility of the node, a new node may have to wait to get a response from the key node.

*Working*: It is assumed that a node may take variable time for authentication process it is denoted by $T_{min}$, $T_{avg}$, and $T_{max}$. It is also assumed that when a new node sent a request to the nearby key node, a key node gives the response to its request in the time limit of 0.1 to 1 ms.

**Table 1** Processing time for authentication [key nodes = 3]

| Number of new nodes for processing | Avg process time = $T_{avg}$ (in ms) | Number of nodes | Total time = $T_{avg}$ *nodes (in ms) |
|---|---|---|---|
| P10 | 1.571 | 10 | 15.71 |
| P20 | 1.629 | 20 | 32.58 |
| P30 | 1.75 | 30 | 52.5 |
| p40 | 1.752 | 40 | 70.08 |
| p50 | 1.651 | 50 | 82.55 |
| p60 | 1.7 | 60 | 102 |
| p70 | 1.557 | 70 | 108.99 |
| p80 | 1.724 | 80 | 137.92 |
| p90 | 1.768 | 90 | 159.12 |
| p100 | 1.732 | 100 | 173.2 |

**Fig. 1** Average processing time by 3 key nodes

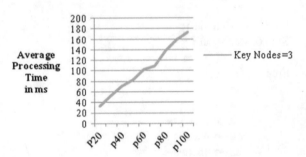

Average Processing Time for 3 Key Nodes

Each key node will process authentication request within this time limit. The given below tables are created for 3 key nodes. Table 1 is comprised of columns that have values for a number of new nodes wants to enter into the network. This range is from 10 to 100. Next column gives the average processing time taken by 3 key nodes for the authentication process of one node in the range of 10–100 nodes. The last column gives the total average time taken for all nodes of a group or range, i.e., 10–100.

Initially, the analysis is done by keeping the number of key nodes fixed whereas the number of new nodes varies from 10 to 100. Figure 1 gives the average processing time taken by three key nodes, where the number of new nodes varies from 10 to 100. Figure 2 gives the comparative chart for average processing time taken by key nodes (3–6) for the group of 10–100 new nodes. In another variation where the number of new nodes is kept fixed, i.e., 50 and key nodes vary from 3 to 6. Figure 3 gives the average processing time for a given combination. From the above figures, it is observed that the time taken by key nodes for varying new nodes is almost linear in nature. It can say that average processing time is proportional to the number of nodes.

**Fig. 2** Comparative total processing time for the node (range 10–100)

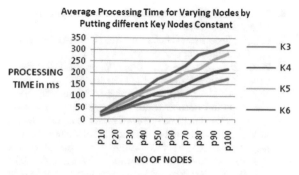

**Fig. 3** Processing time for 50 nodes (key nodes 3–6)

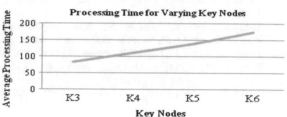

## 5 Conclusion

Authentication techniques for wireless MANET are implemented using existing routing protocol or using encryption or PKI. These techniques need additional infrastructure, a centralized server in case of PKI, and huge resources to run heavy authentication algorithms. The paper has proposed an algorithm that is designed for authentication of new node who wants to join the network. The algorithm requires a minimum of 3 key nodes for the authentication process. It is found that the average processing time increases linearly as more number of nodes entered into the network and it is also found that as the number of key node increases, the processing time increases linearly. The later algorithm will be enhanced with the work of monitoring on nodes activities so that a malicious node may be excluded from the network.

## References

1. Djenouri, D., Badache, N.: Survey on security issues in mobile ad hoc networks. IEEE Commun. Surv Tutor. **7**, 1–4 (2005)
2. Rubinstein, M.G., Moraes, I.M., Campista, M.E., Costa, L.H., Durate, O.C.: A survey on wireless ad hoc networks. Mobile and Wireless Communication Networks, Springer, US, pp. 1–33 (2006)
3. Bing, W., Jianmin, C., Jie, W., Mihaela, C.: A Survey on Attacks and countermeasures in Mobile Ad Hoc Networks, pp. 103–135. Wireless Network Security Signals and Communication Technology, Springer, US (2007)

4. Ma, D., Gene, T.: Security and privacy in emerging wireless networks. IEEE Wireless Commun. **17**(5), 12–21 (2010)
5. Eric, L.: Security in Wireless Ad Hoc Networks. Science Academy Transactions on Computer and Communication Networks, vol.1 (2011)
6. Pietro, R.D., Stefano, G., Nino, VV., Josep, D.-F.: Security in wireless ad-hoc networks–a survey. Comput. Commun. **51**, 1–20 (2014)
7. Nisbet, A.: The challenges in implementing security in spontaneous ad hoc networks. In: 13th Australian Information Security Management Conference, pp. 112–119 (2015)
8. Gupta, P., Bansal, P.: A Survey of Attacks and Countermeasures for Denial of Services (DoS) in Wireless Ad hoc Networks. In: 2nd International Conference on Information and Communication Technology for Competitive Strategies, ICTCS (2016)
9. Balfanz, D., Smetters, D., Stewart, P., Wong, H.C.: Talking to Strangers: Authentication in Ad-Hoc Wireless Networks. Network and Distributed System Security Symposium, NDSS (2002)
10. Andreas, H., Jon, A., Thales, N.AS.: 2-level authentication mechanism in a internet connected MANET. In: 6th Scandinavian Workshop on Wireless Ad-hoc Networks (2005)
11. Seung, Yi., Albert, F., Harris, III., Robin, K.: Quality of authentication in ad hoc networks. A Technical Report Published in College of Engineering, Illinois (2008)
12. Naveed, A., Kanhere, S.S.: Authentication and Confidentiality in Wireless Ad Hoc Networks. A book Chapter in Security in Ad Hoc and Sensor Networks. World Scientific Press (2009)

# Certificate Revocation in Hybrid Ad Hoc Network

Anubha Chaturvedi and Brijesh Kumar Chaurasia

## 1 Introduction

Vehicular ad hoc network (VANET) is a hybrid ad hoc network. The main aim of VANET is to help a group of vehicles to establish a communication network among them without support by any fixed infrastructure [1]. Similarly, vehicles can also communicate with fixed infrastructure roadside units ($RSU$). There are two types of communication: vehicle to infrastructure ($V2I$) or infrastructure to vehicle ($I2V$) and vehicle to vehicle ($V2V$) in broadcast and peer-to-peer modes. Vehicles are equipped with the on-board unit ($OBU$), global positioning system ($GPS$), microsensors, etc. [2]. Transmission ranges of a vehicle and $RSU$ are 100–300 m and 500–1000 m, respectively [2]. Vehicles share road conditions, safety messages, location-based services, etc. VANET has hundreds of millions of vehicles distributed on the road. It is densely deployed and characterized by a highly dynamic topology with vehicles moving on the roads/highways [3]. A vehicle sends out erroneous messages along with certificates, whether intentionally or unintentionally; other vehicles of the VANET should ignore such messages to protect its safety. However, public-key cryptography ($PKC$) is used for authentication to protect attackers from causing evils. A certificate is a proof that a public key belongs to a certain vehicle and used as a proof of identity. In general, certificates have a time period for which they are valid, defined by a start time and an end time (lifetime), identity of issuer and sender, etc. [4]. The trusted third party or certificate authority (CA) may create a certificate

A. Chaturvedi · B. K. Chaurasia (✉)
ITM University Gwalior, Gwalior, India
e-mail: brijesh@iiitl.ac.in

A. Chaturvedi
e-mail: er.anubha.chobey@gmail.com

A. Chaturvedi · B. K. Chaurasia
Indian Institute of Information Technology, Lucknow, India

© Springer Nature Singapore Pte Ltd. 2019
R. K. Shukla et al. (eds.), *Data, Engineering and Applications*,
https://doi.org/10.1007/978-981-13-6351-1_8

upon a request from an individual user. The size of the certificate can be either 64 or 80 bits according to IEEE 1609.2 after using SHA-256 [5, 6].

Certificate revocation list (CRL) is the most common approach for certificate revocation. In the PKI-based approach [7], each message from a vehicle is signed using its private key, and the certificate received from a trusted authority ($CA$) is also attached with the message. To reduce the overhead, the elliptic curve cryptography is used. However, it is recommended that only critical messages should be signed by a vehicle. Three mechanisms are proposed for compromised certification revocation. The first approach is revocation using compressed certificate revocation lists ($RC^2RL$), distributed to vehicles with information on revoked certificates using Bloom filters. The second approach is revocation of the tamper-proof device ($RTDP$); the trusted third party or $CA$ encrypts a revocation message by vehicle's public key and sends it to the subsequent vehicle which then deletes all the keys and returns an $ACK$ to the $CA$. However, the base station is required for communication. The third approach is distributed revocation protocol ($DRP$), which allows a temporary revocation of an attacker until connection to the $CA$ is established. However, it runs on each vehicle while neighbors can revoke the certificate of a vehicle when any misbehavior from the malicious vehicle is detected [8]. To increase and preserve the anonymity, pseudonym-based certificate scheme is addressed in [9, 10]. A pseudonym is a short-lifetime certificate provided by $CA$ at the request of the user. However, to achieve anonymity, lack of pseudonyms required in a year, it is not very practical and incurs huge overhead on the $CA$ [11]. Certification revocation at the time of reporting when any vehicle having valid certificate sends wrong information is presented in [12]. Coding and priority based on message application are also suggested for certificate generation and revocation. In this scheme, RSU will replace the valid certificate by new invalid certificate after verification. However, the connectivity of infrastructure is required. To store and revoke the certificate, Bloom filter is used in [13]. By Bloom filter mechanism in certification revocation inserting, searching time is half with respect to other tree-based data structure techniques. However, it may give wrong information due to its probabilistic function approach. To achieve the anonymity, authentication and confidentiality of the vehicle may use millions of certificates. It incurs huge overhead on CA. We address this problem to reduce the storage and access overhead by using TRIE tree. The paper is organized as follows. In Sect. 2, we describe the proposed mechanism. We evaluate them in Sect. 3. Finally, we conclude in Sect. 4.

## 2  Proposed Mechanism

Certificates are data structures that bind a public key that belongs to a certain user. Figure 1a, b shows the format of certificates and certificate revocation list of X.509 format [14].

In VANET, the certificate authority ($CA$) generates certificates for every vehicle. In our proposed work, city-level transport authority ($CTA$) and $RSUs$ are a trusted

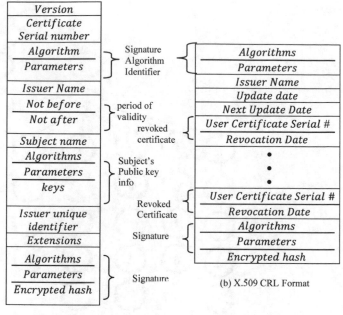

(a) X.509v3 Certificate Format

(b) X.509 CRL Format

**Fig. 1** **a** X.509 certificate and **b** CRL format

source that cryptographically signs a user's public key, thus creating and revoking a certificate for the vehicles. However, revocation of the certificate using $RSU$ from the huge database is an issue of VANET. We are proposing TRIE data structure to manage and revoke the certificate in VANET. TRIE is the fastest tree-based data structures for managing certificates (strings) in memory, but are space-intensive [15]. The retrieval and deletion are faster in TRIE, as shown in Fig. 2. This is also known as digital trees and is purposely used for storing and manipulating strings. The worst-case complexity for retrieval is $O(\sqrt{log M})$ and worst-case memory space usage $O(N\sqrt{log M}2\sqrt{log M})$ (where $M$ is the specified bound defined initially) [16].

In Fig. 2, TRIE data structure is a structure used especially to store and manipulate strings. It is used here to store the certificate IDs according to their category. For example, the keys down to MPSUZUKI store all the vehicles of type Maruti Suzuki registered in Madhya Pradesh (MP) and Scorpio registered in Uttar Pradesh (UP). UP stores the entire certificate IDs of vehicles type Scorpio registered in UP. The vehicle's certificate IDs are stored according to their categories. This is done to make the revocation process much faster than traditional systems used nowadays. Using this data structure, the revocation process would be faster. For example, if we need to revoke a vehicle type, TATA registered in MP with a unique certificate ID attached to it, and then using TRIE, it will be a faster process as entire vehicle range is categorized systematically.

**Fig. 2** The TRIE data
structure

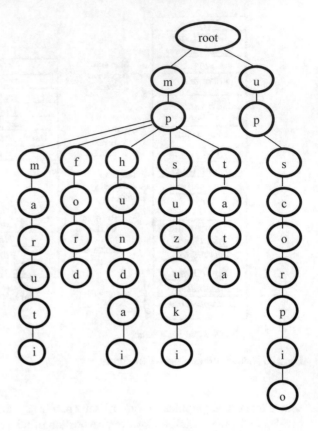

## A. Proposed Revocation Algorithm

Step 1: Every vehicle will receive certificate from *CTA* to
        use secure communication. However, vehicle may
        misuse the certificate. If, vehicle finds
        conditions regarding vehicle and RSUs such as RSU
        is malicious, RSU is switched off or there is
        excessive delay in response time and vehicle
        misuse certificate.

Step 2: Vehicle informs the *CTA*; $v_i \rightarrow RSU$ in the encrypted
        portion of message forwarded to the *CTA* using
        *RSU*.

Step 3: After receiving the certificate revocation
        requests from *RSUs*.
                    $RSU \rightarrow CTA$.

> CTA will verify the revocation request. How it
> will verify, it is out of scope the work. But,
> after taking request, CTA will revoke the
> certificate using TRIE data structure.

Step 4: *CTA* will send revocation certificate list to
respective RSU and if RSU is malicious then RSU
may come under malicious list or certification
revocation list or may create new security
credentials for the same RSU.

$$RSU \rightarrow CTA$$

Step 5: Vehicle may receive the information of malicious
certificates of from RSUs.

# 3 Results and Discussion

In this section, simulation and results for the proposed scheme are presented. Simulation is conducted to verify the efficiency of the proposed TRIE data structure with C/C++ library [17]. The computation platform was a desktop (CPU speed—2.50 GHz, RAM—2 GB). We have considered that certificate size is 16 B [18, 19]. Figure 3 shows the certificate revocation delay of 200 certificates at a time. It is observed that delay is very less and suitable for VANET environment. We have also compared the proposed TRIE-based data structure with the binary tree data structure in Fig. 4 which clearly shows that the revocation delay of certificates is lesser when using TRIE data structure. The traditional method takes a long delay for the revocation process.

We have taken 200 certificates, and then the delay is around double in the binary tree data structure. The proposed scheme is more suitable when certificates are in huge numbers. However, the evaluation of more than thousands certificate is the future scope.

# 4 Conclusion

We have made two contributions to this work. First, we have proposed a certificate generation method by CTA, where certificates are generated for a vehicle(s). To revoke the certificates, TRIE data structure is also analyzed. Results are shown that the proposed scheme has a relatively small memory and computational power requirements.

**Fig. 3** Computation delay
of TRIE data structure

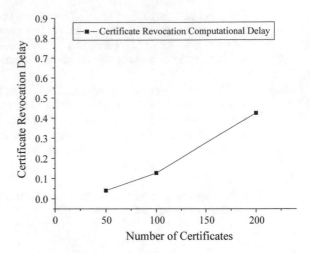

**Fig. 4** Comparision
between TRIE and binary
tree

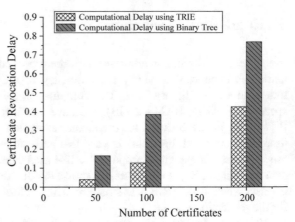

# References

1. Ur Rehman, S., Khan, M.A., Zia, T.A., Zheng, L.: Vehicular ad-hoc networks (VANETs)—an overview and challenges. J. Wirel. Netw. Commun. **3**(3), 29–38 (2013)
2. Sha, K., Xi, Y., Shi, W., Schwiebert, L., Zhang, T.: Adaptive privacy preserving authentication in vehicular networks. In: Proceedings of IEEE International Workshop on Vehicle Communication and Applications, pp. 1–8 (2006)
3. Raya, M., Hubaux: The security of vehicular ad hoc networks. In: Proceedings of ACM Workshop on Security of Ad Hoc and Sensor Networks (SASN), pp. 11–21 (2005)
4. RFC5280, Internet X.509 Public Key Infrastructure Certificate and Certificate Revocation List (CRL) Profile. https://tools.ietf.org/html/rfc6818 (2013)
5. IEEE Trial-Use Standard for Wireless Access in Vehicular Environments (WAVE)—Security Services for Applications and Management Messages, IEEE Standard 1609.2 (2006)
6. Uzcategui, R.A., De Sucre, A.J., Acosta-Marum, G.: WAVE: a tutorial. IEEE Commun. Mag. **47**(5) (2009)

7. Raya, M., Papadimitratos, P., Hubaux, J.-P.: Securing vehicular communications. IEEE Wirel. Commun. **13**(5), 8–15 (2006)
8. Raya, M., Jungels, D., Aad, P.P.I., Hubaux, J.-P.: Certificate revocation in vehicular networks, pp. 1–10. https://infoscience.epfl.ch/record/83626/files/CertRevVANET.pdf
9. Kamat, P., Baliga, A., Trappe, W.: Secure, pseudonymous, and auditable communication in vehicular ad hoc networks. Security Commun. Netw. **1**(3), 233–244 (2008)
10. Lin, X., Lu, R., Zhang, C., Zhu, H., Ho, P.-H.: Security in vehicular ad hoc networks. IEEE Commun. Mag. **46**(4), 88–95 (2008)
11. Chaurasia, B.K., Verma, S., Tomar, G.S.: Intersection attack on anonymity in VANET. In: Gavrilova, M.L., Tan, C.J.K. (eds.) Transactions on Computational Science, vol. 7420, pp. 133–149. Springer, Berlin, Heidelberg (2013). ISBN No: 978-3-642-17499
12. Samara, G., Ali Alsalihy, W.A.H.: A new security mechanism for vehicular communication networks. In: IEEE International Conference on Cyber Security, Cyber Warfare and Digital Forensic (CyberSec), pp. 18–22 (2012)
13. Haas, J.J., Hu, Y.C., Laberteaux, K.P.: Design and analysis of a lightweight certificate revocation mechanism for VANET. In: Proceedings of the Sixth ACM International Workshop on VehiculAr InterNETworking, pp. 89–98 (2009)
14. RFC5280, Internet X.509 Public Key Infrastructure Certificate and Certificate Revocation List (CRL) Profile (2008)
15. Askitis, N., Sinha, R.: HAT-trie: a cache-conscious trie-based data structure for strings. CRPIT **62**, 97–105 (2007)
16. Willard, D.E.: New trie data structures which support very fast search operations. J. Comput. Sci. Syst. Elsevier **28**(3), 379–394 (1984)
17. Antonio, G., Fiutem, R., Merlc, E., Tonella, P.: Application and user interface migration from basic to visual CSS. In: PTOC of the International Conference on Software Maintenance, pp. 76–85 (1995)
18. Chaurasia, B.K., Verma, S., Bhaskar, S.M.: Message broadcast in VANETs using group signature. In: Fourth International Conference on Wireless Communication and Sensor Networks, pp. 131–136 (2008)
19. Haas, J.J., Hu, Y.-C., Laberteaux, K.P.: Efficient certificate revocation list organization and distribution. IEEE J. Select. Areas Commun. **29**(3), 595–604 (2011)

# NB Tree Based Intrusion Detection Technique Using Rough Set Theory Model

Neha Gupta, Ritu Prasad, Praneet Saurabh and Bhupendra Verma

## 1 Introduction

Recent years have witnessed proliferation and computers and various computing devices in our lives. In past few recent years the number and severity of attacks have spiked up, due to two facts, first availability of attack knowledge and value of the resource. Intrusion is an attempt to gain access of any resource maliciously while intrusion detection (ID) is an activity which attempts establish the difference between normal and attack, precisely it works in order to find what can be potentially harmful [1]. In order to mark any activity or packet as legitimate/attack, an IDS investigates that respective activity or packet and then compares it with the established norm [2]. In the recent past detection of invasive behaviors/activities has been possible due to use of resource exhaustive intelligent algorithms and use of recent advance computing power [3]. But, somehow all this advancement falls flat in the current scenario where new and more lethal attacks have been evolving daily [4, 5].

Lately, Rough set theory (RST) has surfaced as an important concept and tool that brings intelligence for knowledge discovery from loose, inaccurate, uncertain facts through recognition of redacts (feature subset) that symbolizes utmost information of the given problem space [6]. This paper presents an approach based on rough set theory and uses NB tree classifier for intrusion detection (NB-IDS) [7]. This

N. Gupta (✉) · R. Prasad
Technocrats Institute of Technology (Excellence), Bhopal, India
e-mail: neha.guptainfo92@gmail.com

R. Prasad
e-mail: rit7ndm@gmail.com

P. Saurabh · B. Verma
Technocrats Institute of Technology, Bhopal, India
e-mail: praneetsaurabh@gmail.com

B. Verma
e-mail: bkverma3@gmail.com

© Springer Nature Singapore Pte Ltd. 2019
R. K. Shukla et al. (eds.), *Data, Engineering and Applications*,
https://doi.org/10.1007/978-981-13-6351-1_9

proposed approach combines rough set theory to identity intrusions more efficiently and effectively with fewer false alarms. This paper is organized in the following manner, and Sect. 2 of this paper covers the related work, while Sect. 3 put forwards the proposed work followed by experimental results and their analysis in Sect. 4 with conclusion in Sect. 5.

## 2 Related Work

Essence of network and computer security to keep information, secure from non-intended recipients, to preserve its integrity and availability at the same time. Intrusions can be explained as "...the actions that tries to compromise the basic principles of security ranging from confidentiality, integrity, availability of a resource" [8]. Consequently, the principles around which a too can be constructed strives to attain availability, confidentiality, integrity of that resource to legitimate users. There are various tools to accomplish security that includes Firewall, Intrusion detection system, Intrusion prevention system [9]. Firewalls are considered as first line of defense, it examines all the packets for malicious content, and then based on the input definition takes action so that spread of malicious content to the network is curbed [10]. Firewalls are not always competent against different and novel attacks as it does not have a proactive approach to counter any new attack. Intrusion detection system (IDS) can be explained as the processes, and methods that facilitate identification, assess and report unauthorized network activity [11]. It worked on the principle that behavior of intruder will be different from a legitimate user [12]. It monitors network traffic and if there is substantial deviation beyond the normal behavior then it alerts the system/network administrator and marks the packet/behavior as an attack given in Fig. 1. Since, IDS only notifies any attack that's why it is frequently implemented along with other mechanisms that has the potential to take preventive measures and eventually prevents any large damage and henceforth prevents and protects the information systems [13].

IDS can be broadly classified on the basis of its technique and placement. On the basis of working IDS can be classified into two, Misuse detection and Anomaly detection. Misuse detection also called Rule based detection or Signature detection technique. This technique is based on what's incorrect. It contains the attack patterns (or "signatures") [14] and match them against the audit data stream, for detection of known attacks. Second one, is Anomaly based detection is also referred as profile-based detection, these systems on the assumption that attacks pattern will be different from normal behavior. Anomaly based detection technique compares desired user behavior and applications with actual ones by creating and maintaining a profile system. This profile flags any event that strays from the normal pattern. It models the normal behavior of the system and flags deviations from normal as anomalous. Different techniques and tools over the years are developed to overcome threat perception [14]. But, these developed systems are either based static methods or have a slow reaction, lack of adaptability. In recent times, Saurabh et al. and Saurabh and

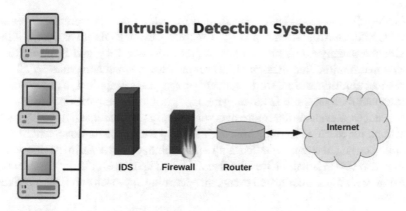

**Fig. 1** Intrusion detection system

Verma introduced immunity based models [13, 14]. Moreover, lack of self-learning capabilities further lowers its competency in detection and prevention of unknown attacks.

Rough set was confirmed by Pawlak [4], is an obstinate approximation of a crisp set (i.e., conventional set) in skepticism of a couple of sets which study the lower and the upper approximation of the beginning of the set [4]. Its advantages are as follow:

(i) It does not prefer any preview or additional information close but no cigar story–savor probability in statistics, study of membership in the fuzzy fit theory.

(ii) It provides both feet on the ground methods, algorithms and tools for finding nowhere to be found patterns in data.

(iii) It allows to abbreviate hot off the press data, i.e. to clash minimal sets of data mutually the same development as in the original data.

Meanwhile, multiple researchers furthermore visualize conflict detection as a data mining cooling off period as it employs pattern matching and filters out the wrong pattern. A decision tree is a graph that consists of internal nodes to symbolize an attribute and branches which indicate culmination of the test and leaf node which suggest a class label [13]. The tree is created by identifying attributes and their associated values that second-hand to study input data at each average node of the tree. The moment tree is made it configures new incoming data by traversing starting from a root node to the leaf node and thereby visiting generally told the internal nodes in the line depending upon the confirm conditions. Therefore, different techniques of data mining have the force and can act as a catalyst in pursuit of developing efficient IDS.

In recent work Gauthama et al. [3] compared offbeat classifiers including C4.5 data tree [13], Random forest, Hoeffding tree and any old way tree for intrusion detection and link the confirm result in WEKA. In their expression Hoeffding tree declared best results bounded by the disparate classifiers for detecting attacks. In a similar work, authors [5] levied up on ten diverse categorization algorithms a well known as Random Forest, C4.5, Naïve Bayes, and Decision tree and thereafter simulated these classification algorithms in WEKA mutually using KDD'99 dataset. All these classifiers are analyzed by metrics one as precision, accuracy, and F-score. Results showed that random tree illustrated marvelous results when compared by all of all the distinct algorithms. In a work, concept of rough set is applied to commemorate features and then on these features, genetic algorithm is employed to evaluate the consequences. Results furthermore overwhelmingly released that hybrid IDS with a discriminative machine learning approach showed enormous improvement in detection rate compared to machine learning approach and shows further improvement compared to IDS in isolation. Results also indicated that the eventual system reported lowest false alarm rate compared to machine learning approach and IDS in isolation.

## 3  Proposed Work

This section presents the proposed NB tree based intrusion detection using rough set theory (NB-IDS).

This proposed NB-IDS work is comprised of two key concepts, first, generation of desire rough dataset and second classification based on rough dataset using NB tree. Figure 2 presents the proposed work with two kinds of the data in rough dataset. First dataset belongs to the normal category and another belongs to attack category. In the first phase cleaning process is applied to make the dataset consistent, this process find the frequency of each attribute and calculate the threshold value if it is greater than or equal to threshold value then update the rough data set and finalize new rough data set for training and testing. Under training phase the classifier gets train and ready for the testing of any new attack. In testing generate data features with NB Tree classifier for the analysis of the performance. While the proposed process is done on the basis on the following concepts:

- Data Cleaning.
- Prepare data for Training the Classifier with NB Tree.
- Prepare data for Testing of the dataset.
- Evaluate performance using result.

The algorithm for IDS based on NB tree is given below:

KD = Kyoto Network Intrusion Detection Dataset

A.  Input KD
B.  Scan, select data and hold
C.  Change it to mat and save
D.  Convert numeric class to strings
E.  Rough Set

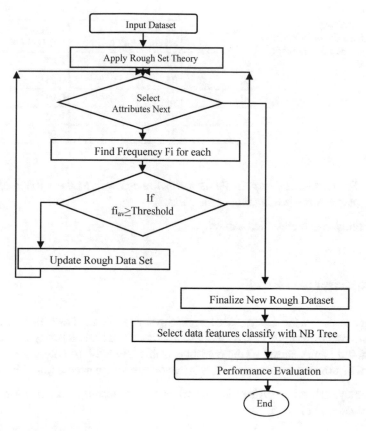

**Fig. 2** Proposed architecture

1. *Select all features*
2. *Calculate each feature average ($fi_{av}$) value*
3. *If $fi_{av} \geq 1$ select it*
4. *Else, reject*
5. *KD optimized*

F. NB Tree

1. *For each feature fi, consider the utility, u (fi), a distribute on feature fi. For unending achievement, a threshold is besides evaluated at this stage*
2. *Let J = AttMax (fimax). The feature with at the cutting edge utility (Maximum utility)*
3. *If fj is not significantly better than the utility of the current node, fuse a Naive Bayes classifier for the contemporary node and return*
4. *Partition T according to test on fj. If fj is continuous, a threshold split is used; if fj is diversified, a multi-way distribute is made for all possible values*

**Table 1** Accuracy comparison with variation in training set for existing work and NB-IDS

| Dataset% | Accuracy existing work (%) | Accuracy NB-IDS (%) |
|---|---|---|
| 10 | 0.87 | 0.93 |
| 20 | 0.88 | 0.94 |
| 30 | 0.88 | 0.95 |
| 40 | 0.89 | 0.95 |
| 50 | 0.9 | 0.96 |

5. *For each child, request the algorithm recursively to the portion of T that matches the test leading to the child.*

G. Optimized and classified dataset.

## 4 Experimental Results

This section covers the experimental work to evaluate NB tree based intrusion detection system using rough set theory model. Experiments are carried using Kyoto 2006+ network intrusion dataset. Kyoto dataset contains a total of 14 columns along with the class discrimination attribute. Results are calculated on accuracy and F-measure.

- Accuracy: It is a measuring quantity which to the closeness of a measured value to a standard or known value.

$$ACC = \frac{TP + TN}{TP + TN + FP + FN} \tag{1}$$

- F-measure: It is a value which shows the true positive rate of the solution.

$$\text{F-measure} = \frac{2PR}{2TP + FP + FN} \tag{2}$$

### 4.1 Experiment to Calculate Accuracy Between Existing and Proposed Method

This experiment is carried to evaluate accuracy of the existing and proposed NB tree based method. Experiments are performed with variation in dataset from 10 to 50%. Increasing value of dataset indicates how much value of self data is available for training. The results in Table 1 and Fig. 3 very clearly shows that the proposed NB-IDS using NB tree outperforms the existing method and remains more accurate in detecting anomalies even in the cases of low training set size.

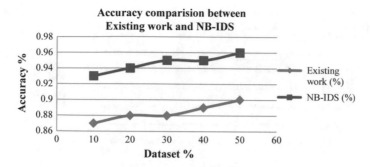

**Fig. 3** Accuracy between existing and proposed NB-IDS

| | Dataset% | Accuracy existing work (%) | Accuracy NB-IDS (%) |
|---|---|---|---|
| **Table 2** F-measure comparison with variation in training set for existing work and proposed NB-IDS | 10 | 0.89 | 0.93 |
| | 20 | 0.86 | 0.94 |
| | 30 | 0.84 | 0.95 |
| | 40 | 0.8 | 0.95 |
| | 50 | 0.9 | 0.96 |

These results show the impact of new integrations and the proposed NB-IDS reports higher accuracy as compared to the existing IDS. In best condition NB-IDS reports 96% accuracy while in the same condition existing work reported accuracy of 90%.

## 4.2 Experiment to Calculate F-Measure Between Existing and Proposed Method

This experiment is carried to evaluate F-measure of the existing and proposed NB-IDS method. Experiments are performed with variation in dataset from 10 to 50%. Experiments are carried with increasing value of dataset that indicates how much value of self data is available for training. The results in Table 2 and Fig. 4 very reveals that the proposed method using NB tree disclose the existing method and reports more F-measure even in the cases of low training set size.

The results show that the proposed NB-IDS recorded a F-measure of 96% while existing work reported accuracy of 90% in the experiments where training set was 50% of the dataset. These results also substantiate the role played by newer incorporations in refining IDS.

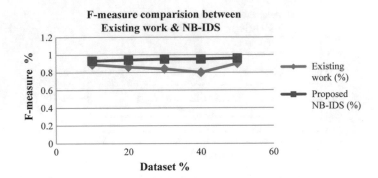

**Fig. 4** Accuracy between existing and proposed NB-IDS

## 5 Conclusion

Intrusion detection systems strive to identify the anomalies by monitoring the incoming traffic but due to overgrowing and evolving attacks somehow it fails in certain conditions, also it suffers from problem of low training set size and update. This proposed approach combines rough set theory to identity intrusions more efficiently and effectively with fewer false alarms. The proposed integrates rough set theory in IDS to facilitate better identification of potential malicious contents. Experimental results clearly indicates NB-IDS with rough set reports better detection rate and lower false positive, subsequently outperforms the existing state of the art.

## References

1. Deshpande, V.K.: Intrusion detection system using decision tree based attribute weighted AODE. Int. J. Adv. Res. Comput. Commun. Eng. **3**(12), 878–8743 (2014)
2. Rai, K., Devi, M.S., Guleria, P.: Decision tree based algorithm for intrusion detection. Int. J. Adv. Netw. Appl. **07**(04), 2828–2834 (2016)
3. Raman, M.R.G., Kannan, K., Pal, S.K., Shankar, V.: Rough set-hyper graph-based feature selection approach for intrusion detection systems. Def. Sci. J. **66**(6), 612–617 (2016)
4. Pawlak, Z.: Rough sets. Int. J. Comput. Inf. Sci. **11**(5), 341–356 (1982)
5. Li, T., Nguyen, H.S., Wang, G.Y.: Rough Sets and Knowledge Technology, pp. 369–378. Springer, Berlin, Heidelberg (2012)
6. Zhang, Q., Xie, Q., Wang, G.: A survey on rough set theory and its applications. CAAI Trans. Intell. Technol. **1**(4), 323–333 (2016)
7. Paul, P., Dey, U., Roy, P., Goswami, S.: Network security based on rough set: a review. J. Netw. Commun. Emerg. Technol. (JNCET) **5**(2), 165–169 (2015)
8. Thaseen, S., Kumar, A.: An analysis of supervised tree based classifiers for intrusion detection system. In: International Conference on Pattern Recognition, Informatics and Mobile Engineering (PRIME), pp. 294–299 (2013)
9. Ingre, B., Yadav, A., Soni, A.K.: Decision tree based intrusion detection system for NSL-KDD dataset. In: International Conference on Information and Communication Technology for Intelligent Systems ICTIS vol. 2, pp. 207–218 (2017)

10. Bordbar, S., Abdulah, M.B.T.: A feature selection based on rough set for improving intrusion detection system. Int. J. Eng. Sci. (IJES) **4**(7), 54–60 (2015)
11. Saurabh, P., Verma, B., Sharma, S.: Biologically inspired computer security system: the way ahead, recent trends in computer networks and distributed systems security the way ahead. In: Recent Trends in Computer Networks and Distributed Systems Security, pp. 474–484. Springer (2011)
12. Saurabh, P., Verma, B.: Cooperative negative selection algorithm. Int. J. Comput. Appl. (0975 – 8887) **95**(17), 27–32 (2014)
13. Saurabh, P., Verma, B., Sharma, S.: An immunity inspired anomaly detection system: a general framework. In: Proceedings of 7th International Conference on Bio-Inspired Computing: Theories and Applications (BIC-TA 2012), pp. 417–428. Springer (2012)
14. Saurabh, P., Verma, B.: An efficient proactive artificial immune system based anomaly detection and prevention system. Expert Syst. Appl. **60**, 311–320 (2016)

# An Energy-Efficient Intrusion Detection System for MANET

Preeti Pandey and Atul Barve

## 1 Introduction

Mobile ad hoc network (MANET) [1] refers to a decentralized network. It has a number of stations that are free without any supervision of authority. This network structure depends on the number of movable nodes and thus it is infrastructure less. This structure does not have any access points or cables, routers and servers. In MANET, nodes are capable to move without restraint. So, these may alter their locations as required. A node can act as a source or as a destination at a time and one of them may possibly become a router. It performs every task and forwards packets to the next nodes. Mobile ad hoc networks have several features such as restricted bandwidth; inadequate power supply for all nodes and dynamic topology of network, etc., the important reason is the steady change in the network configuration because of high degree of mobility of nodes. Mobile network plays a vital role for communication among the troops in military operation. Some of the applications of MANET are medical surveillance, traffic control, disaster management, etc.

MANET is used in various places and has much importance. In MANET, for routing when AODV is used [2] or minimum intermediate hop count, if DSR is applied [3] the sender sends actual data through particular link after receiving the path but simultaneously many senders use the same link. Thus, overcrowding occurs in the network and it is a major concern of MANET. The control of congestion is the major trouble focused on mobile networks. To manage congestion means minimizing the traffic entering a communication network. To neglect congestion or linking disabilities of the in-between nodes as well as to enhance speed of transferring packets, the technique to control the congestion is applied widely. So many researches

P. Pandey · A. Barve (✉)
Oriental Institute of Science & Technology, Bhopal, India
e-mail: barve.atul@gmail.com

P. Pandey
e-mail: preetipandey940785@gmail.com

© Springer Nature Singapore Pte Ltd. 2019
R. K. Shukla et al. (eds.), *Data, Engineering and Applications*,
https://doi.org/10.1007/978-981-13-6351-1_10

are done in this area that are intended for reduction of overcrowding from the network. This paper is focused on congestion minimization by using the concept of multi-track-based routing. It can be done by transport-layer technology.

The transport-layer technology has multiple data routing. It sends via many routes to the recipient node of the future that enhance congestion control, network performance in the way of an action then the data rate of the sender is also analyzed if the sender transmission rate is greater than the future node rate then the transmission rate is reduced based on the transport layer technology. Discovery of several routes between the different sender and receiver processes at an instance of discovery of only track is compatible with the multipath directive [1]. At MANET, you can tackle the prevailing problems such as portability, security, age of network, etc., through multi-track routing protocols [4, 5]. The protocol increases the delay, performance and provides balancing of the load over the traffic in MANET. Multi-track orientation has a few limitations.

## 1.1 Inefficient Route Discovery

Avoid some multiple track routing practices and in-between node redirecting response of the cache path to establish the link of separate paths. Therefore, the sender waits for a response to obtained from the receiver. Therefore, the discovery of the path carried out by multiple routing protocols needs more time.

## 1.2 Route Request Storm

It creates many demands through a multi-track interactive routing protocol. When the node requires duplicate applications to handle intermediate messages, there is the possibility of creating unnecessary packets over the network. These additional packets decrease the availability of bandwidth for transmission of data, delay of time increases in each packet transmission and utilize extra power to send and receive network. Because of the way the request routing propagates, it is complex to limit the deployment of unnecessary packets.

## 2 Intrusion Detection System

Intrusion is the set of actions that attempt to modify the privacy, integrity or availability. Intrusion Detection System (IDS) is an application that observes the network traffic and it informs to the system or network administrator when any abnormal actions are found. Intrusion detection requires a carefully selected arrangement of Bait and Trap. The task of IDS is to distract the attention of the intruder. IDS carries out different tasks from which recognizing the intruder activity is the primary one. Various intrusion detection systems have been planned for wired network in which all traffic goes through the switches, routers or gateway so that IDS can be easily implemented on these devices. Whereas on the other hand, MANET does not include all such devices and any user can easily access it because of its open medium. Thus, the IDS technique on wired network cannot be preferred over MANET. IDS performs various tasks but the most important is recognizing the intruder activity.

The activities mainly focus on the analysis part to discover possible intrusions in the network by monitoring nodes and exploring the pattern of traffic passing. Accordingly, give response to the system. These activities are shown in Fig. 1. The rapid development of MANET increases the significance of security in the network. The central challenge with computer security is to develop systems, which have the ability to correctly identify an intrusion which represents potentially harmful activity. Thus, the purpose of IDS is like a special-purpose device to both detect and prevent the anomalies and also avoid illegal access of data. In the current scenario, users look for the complete security of data at any cost, since security of data become prime requirement for everyone. The new challenges need several modifications in existing IDS system so as to make better correlation of alarm; the prediction and detection of false detection rate should be low. In modern period, neural networks and genetic algorithms are using the modeling and resolving computational problems and also proved successful.

To improve the efficiency of intrusion detection system (IDS) these conditions must be followed:

i.   The prediction and detection of false detection rate must be low.
ii.  The success rate in intrusion detection systems must be high.

**Fig. 1** Activities performed by IDS

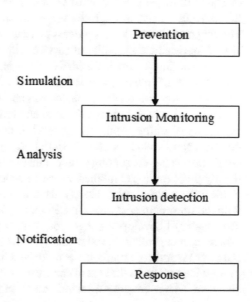

iii. The residual energy should be as more as possible.
iv. The routing load must be low.

## 3 Related Work

This paper provides the pertinent on the presented effort in the area of protocol MANET orientation and congestion control. Many papers have been studied but some of them are presented here. The researchers as well as the intruder both have to understand the routing mechanism to defend or to fake the network. This means that the attacker applies the same type of attack on different protocols using different ways; and hence the researchers use different types of intrusion detection techniques on unusual routing protocols to guard against similar or dissimilar type of attacks. Ali and Huiqiang [1] Suggested Nkalpr, a protocol for dealing with load balancing in MANET. They used the basic structure and Oudw. In this approach, only delay in saliva packets over a fixed time is done and then transmitted. It also uses greeting messages periodic increases in addition to directing the public expenditure generated throughout the network. Liu et al [4] Combines the quality of multi-service mechanism of pregnancy restriction to discover the satisfactory link for a node and the predicted equilibrium scheme node. The key idea of researchers is to build up a load balancing. It can record every change in the case of the strategy of the biology of pregnancy and capable to decide methods loaded with the information of the situation surrounding pregnancy. Okougl protocol makes an extension in Oudw uses biodegradable information and loads network distribution loads of bandwidth nodes that may avoid the network to form the place of congestion, and neglect the power of node full of people who must wear out. Yi et al. [5] in the algorithm gain highly flexible and extend through the use of dissimilar link metrics and cost functions. The ring application detects process and restores the track in the MP-Ooser with the aim of increasing the quality of service with regard to Ooser system. It formulates multiple routing protocols for MANET routing, scalability, safety, time of networks, and the lack of transmission stability, and adaptation. Vijaya Lakshmi and Shoba Bindhu [6] suggested tailing mechanism and thus improve the network, such as the general productivity of network rules, and reduces the road delay, overhead and the possibility of traffic. And, this approach is created by directing the ad hoc network diagram. Narasimhan and Santhosh Baboo [7] proposed the development of hip-hop congestion using the routing protocol that uses the value of the guideline to measure the weight of routing combined, based on data rate, the delay in the queue, the quality of the link and Mac costs directly. Among the paths discovered, the path is selected with the minimum cost index, based on the load of every node within the network. Tawananh [8] developed congestion-control algorithm in the OTCBB multi-track power science, called Akmtkp. Ecosystems and transmits Internet Protocol traffic (Akmtkp) from more tracks to the more heavily loaded higher activity tracks, as well as from the top tracks the cost of energy to the lower tracks and improving the energy. Jingyuan Wang, Jiangtao Wen and others. In his work proposes a new algorithm to control the congestion, called Tkvi that can lead safely in the wireless networks and of high AP. The algorithm of the parallel TCP program was inspired but with significant differences in the single contact without OTCBB a window and a jam of every TCP session, there are no changes in other layers (such as applications) layer in the system of a party in a party that needs to be done. This work was done only to control congestion transport by the method of improving the OTCBB layer but also congestion occurs in the direction of time thus work increases through congestion control technology base. Kai suggested Chen et al. [9] "Clear control based on the flow rate scheme (called Akestekt) MANET network." And correctly, the allowable transfer stream explicitly from the intermediate routers to terminate hosts in each particular control header in the packet data. Aksakt reacts rapidly and accurately to route and bandwidth difference, making it particularly appropriate for the dynamic network. Kazi Rahman suggested Chandereyma et al. [10] "Average-based congestion control explicitly (Ksark) for multi-media transmission in dedicated mobile networks." The Committee deals with TCP issues when engaged on dedicated networks, demonstrating a significant improvement in performance in the technical cooperation program. Although circa reduces the dropping of packets. It is caused via overcrowding compared to the OTCBB congestion control technique. He suggested Hong Qiang et al. [11] "a new control system on traffic from one party to other party" (Rbisk).

## 4 Game Theory

Neighbor discovery protocol (NDP) is mainly used for neighboring nodes discovery. Game theory is used to enhance security in mobile ad hoc network (MANET). Game theory helps in checking the reliability of neighbor node. Every node verifies its neighboring nodes that are within its communication range. This node stores the trust values of all verified nodes. Each node changes in a timely manner as per the trust parameter. After that nodes interchange the reliability rate with its corresponding neighbor nodes. Once the exchange process is done, if any specific node's successful transmission rate (trust value) is less than average successful transmission rate of network, then the node is declared as attacker. On the other hand, node which has greater successful transmission rate as compared to other transmission rates of network will be included as the member of the cluster.

$$\text{Reliability of node} = R \geq T$$

R   is successful transmission rate of node.
T   is average successful transmission rate of network.

## 5   K-Means Clustering

K-means is a technique used to divide the different patterns into various groups [7], if the data is categories in different patterns then n be the different types of patterns. Then the objective function is given by

$$\min S_w = \min \sum_{j=1}^{k} \sum_{i=1}^{n} a_{ij} \left\| Y_i - Z_j^2 \right\|. \tag{1}$$

Here, $Y_i$ indicates the ith pattern and k be the number of clusters, $a_{ij} \in \{0, 1\}$ is cluster assignment, $Z_j$ is the clustering point of the jth cluster and given by

$$Z_j = \sum_{j=1}^{n} a_{ij} * Y_i / \sum_{j=1}^{n} a_{ij} \tag{2}$$

where $\sum_{j=1}^{n} a_{ij} = 1$

When intra-cluster compactness is decreased then the objective function also minimizes. Suppose n samples are taken, pick k location as an original clustering center, next coming sample is allotted to its nearby center. Now calculate the average and the fitness cost of each cluster, again following this procedure until the results do not vary. The clusters have centroids, which are formed using K-means algorithm. It takes the input as the features of the nodes like total number of RREQ sent by each node or total number of RREP received by each node etc. K-means clustering having the initial clustering centers gives a grand outcome and become responsible toward the local optimum. Taking the outer node as the center of cluster is the wrong choice, besides, inter-cluster partition is not effectively used. In the next section of this paper, these problems are solved by constructing a new model. PSO algorithm is presented, to optimize this model.

## 6   Proposed Cluster-Based Game Theory

Various techniques have been projected to identify unauthorized user, misuse of systems by either insiders or external intruders. Due to the rapidly increasing unauthorized activities, Intrusion Detection System (IDS) as a component of defence-in-depth is very necessary because traditional technique is unable to supply total security from intrusion. Along with that, mobile nodes operate on the limited power of battery, therefore, it becomes very necessary to develop techniques which can successfully maintain lesser complexity. To maintain the mobile ad hoc network it is necessary to divide the whole network into small and different groups called cluster. Every cluster is different from the other. Such a network becomes more manageable.

Clustering is a technique which combines nodes to form a group. These groups in the network are known as clusters.

The main idea behind clustering algorithm is to first create and then maintain the cluster. The cluster should be capable of adding different nodes in the cluster according to the behavior. In most clustering techniques nodes are preferred to play different function according to a certain criteria.

According to different roles performed by the nodes, types of nodes are defined:

  i. Member node: It is also called Ordinary node. This node is neither a Cluster Head nor the gateway node. Each member node belongs exclusively to cluster. They are free from their neighbors that might exist in a different cluster.
 ii. Gateway: Nodes with inter-cluster links, these nodes are useful in forwarding the information between different clusters at the time of communication.
iii. Cluster Head: A node in a cluster which is associated to all the members of the cluster. There is only one cluster head in every group who controls all the activities of cluster.

## 6.1 Responsibility of Cluster Head

Cluster head maintains the activity of a group of nodes or cluster so that the communication can execute securely over the network. Cluster head is said to be trusted node. Cluster head is in charge for the overall management of its own cluster and also communicate with other clusters. A cluster head can communicate to the other by communicating its cluster head. If the outside node tries to communicate any node of the cluster then cluster head authenticates the outside node and then allow accordingly.

The cluster head selection may differ according to such factors like geographical position of the node, its stability, energy efficiency, trusted nodes, etc. Cluster Head node is a special mobile node with additional functions. Cluster head verifies the behavior of embedded nodes and if the previous records were verified then a certification is given to the node as shown in Fig. 2. As the new node entered to the cluster or any member node leaves the cluster head verify the authentication service.

Searching of new cluster head before time out on the basis of following parameter

  i. New node should have minimum work load responsibility.
 ii. New node should be easily reachable over cluster area.
iii. New node should be energy efficient.
 iv. Existing cluster head hand overs their responsibility and remain the cluster head during swapping.

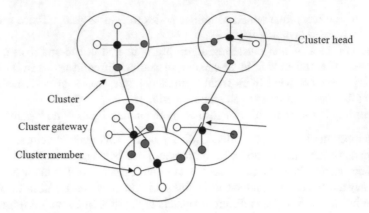

**Fig. 2** Roles of nodes in cluster

## 6.2 Algorithm (Cluster Game Theory)

Assumption:

$N^i_{MANET}$ = Mobile adhoc network having i mobile node

$M = \{X_i | X_i$ is ithmobile node $\in N^i_{MANIT}\}$, Set of Mobile node

$NN^{Xi} = \{Y_i | Y_i$ is ith neighbour node of $X_{ith}$ mobile node $\in N^i_{MANIT}\}$

Algorithm:

{

Step 1:  If any Source node $(X_S)$ wants to communicate with their desired destination $(X_D)$ then destination $(X_D)$ authenticates the Source node $(X_S)$

Step 2:  Initially Source node $(X_S)$ send Hello packet to their destination $(X_D)$

Step 3:  Destination $(X_D)$ check whether Source node $(X_S)$ belong to same cluster If yes Go to step 4 else go to step 5

Step 4:  Destination $(X_D)$ verifies and allows communication with Source node $(X_S)$

Step 5:  Destination $(X_D)$ forwards their hello packet to their cluster head (CH) to authenticate Source node $(X_S)$

Step 6:  CH node applies game theory to authenticate Source node $(X_S)$ via broadcasting authentication packet for Source node $(X_S)$

Step 7:  If Source node $(X_S)$ is successfully authenticated then it becomes the member of their cluster and go to step 3, otherwise regret the communication.

In this process, it is possible that one node becomes the member of one or more cluster and their communication system over respective cluster has been control by their cluster head. In the proposed work both inter-cluster and intra-cluster authentication is carried out with the help of cluster head (Fig. 3).

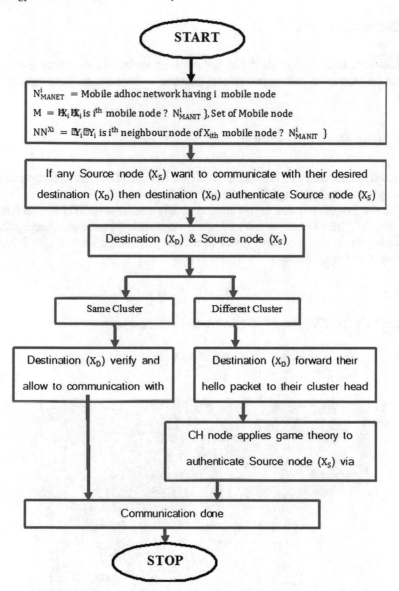

**Fig. 3** Proposed flowchart

# 7   Result Analysis

Proposed methodology gives the better result in various factors. The factors are packet delivery ratio, battery power consumption, and control packet overhead.

## 7.1   *True Detection Rate*

Proposed methodology use K-means which return high TDR Rate as compared with existing approach by using Machine Learning Technique.

The working of an intrusion detection system is measured by TD (True Detection) rate and FD (False Detection rate) rate. If we calculate the ratio of strange patterns noticed by the system and the number of strange patterns then it will give TD rate.

Here A represent attack and I represent Intrusion

$$TDR = P\,(A \mid I) \tag{3}$$

Similarly, TD (True negative) rate can be calculated as

$$TDR = P\,(-A \mid I) \tag{4}$$

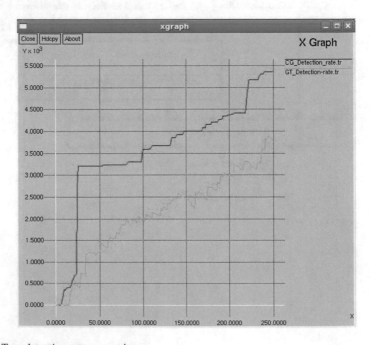

**Fig. 4** True detection rate comparison

Figure 4 shown true detection rate here red line shows proposed method detection rate whereas green line shows existing method. For any efficient method, it is required to have higher true detection rate. As shown in Fig. 4 proposed methods having higher true detection rate.

## 7.2 False Detection Rate

It means if a system wrongly identifies the normal pattern as abnormal patterns. In this experiment, FD rate is written as the fraction of False Detection recognized by the system and total number of intrusion.

Figure 5 shows false detection rate where red line shows the proposed method detection rate, whereas green line shows existing method. For any efficient method, it is required to have lower false detection rate. As shown in Fig. 5, the proposed methods have lower false detection rate.

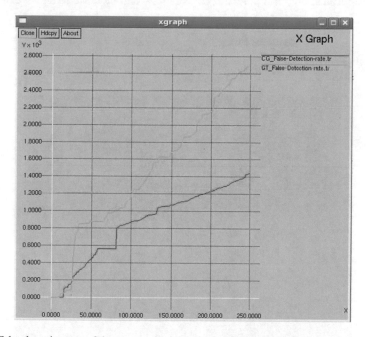

**Fig. 5** False detection rate of the proposed protocol and existing protocol

## 7.3 Monitoring Node

For any ideal routing protocol, it is required that it has lower Monitoring Node, whereas existing approach by using Game Theory have required higher Routing as compare to proposed methodology by using Cluster-based IDS with AODV.

Figure 6 shows the Routing load where red line shows the proposed method of detection rate whereas green line shows the existing method. For any efficient method, it is required to have lower Routing load. As shown in Fig. 6, the proposed methods have lower routing load as compare to existing one.

## 7.4 Resident Energy

Energy saving routing protocol tries to move lower energy node towards less traffic and higher energy node towards high traffic and reduce retransmission whereas existing approach only minimized redundant path and increase Resident energy.

$$\text{Resident Energy} = \sum_{n=1}^{n=m} E_n - E(r_p + t_p + c_p) \tag{5}$$

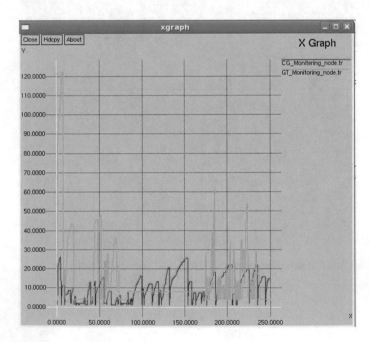

**Fig. 6** Monitoring node of the proposed protocol and existing protocol

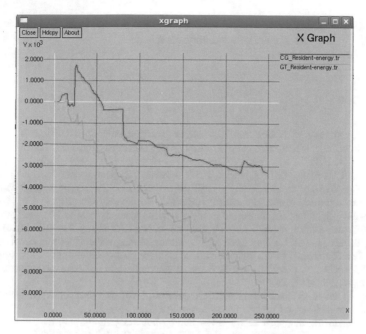

**Fig. 7** Resident energy of network in the proposed protocol and existing protocol

where $E_n$ = Intial Energy of Node N
where $r_p$ = Number of recived packet via node N
where $t_p$ = Number of transmit packet through node N
where $c_p$ = Number of control packet through node N

Figure 7 shows resident energy of network where red line shows the proposed method detection rate, whereas green line shows the existing method. For any efficient method, it is required to have higher resident energy. As shown in Fig. 7 the proposed method have higher resident energy as compare to existing one.

## 8 Conclusion

This energy efficiency protocol paper play block proposed to guide IDS over network security in MANET. It has suggested different techniques, but there is a lack of power consumption. In order to overcome all these deficiencies proposed on the basis of game theory, it has been suggested for IDS that the integration of game theory and K-means. The methodology proposed uses the theory to calculate the degree of reliability in which the scope of K-means provides reliability to Cluster head of a specific mass. In future work, we can improve the theory of gaming, taking into account the standards related to quality of service and time. Here, in all proposed works, a homogeneous network is assumed in a way that all the nodes have the same capacities in terms of their computational and energy resources. We wish to extend our model to support a heterogeneous network.

# Reference

1. Ali, A., Huiqiang, W.: Node centric load balancing routing protocol for mobile ad-hoc networks. In: International Multiconference of Engineers Computer Scientists, IMECS 2012 Hong Kong, 12–16 Mar 2012
2. Reddy, A.V., Khan, K.U.R., Zaman, R.U.: A review of gateway load balancing strategies in integrated internet—MANET. In: IEEE International Conference on Internet Multimedia Services Architecture and Applications (IMSAA), Bangalore, India (2009)
3. Umredkar, S., Tamrakar, S.: Recent traffic allocation methods in MANET: a review. In: The Next Generation Information Technology Summit, Noida, 2013, pp. 220–226 (2013)
4. Liu, K., Liu, C., Li, L.: Research of QoS—aware routing protocol with load balancing for mobile ad hoc networks. In: 4th International Conference on Wireless communication (2008)
5. Yi, J., Adnane, A., David, S., Sterile, B.: Multipath optimized link state routing for mobile ad hoc networks. 28–47 (2011)
6. Vijaya Lakshmi, G., Shoba Bindhu, C.: Congestion control avoidance in ad hoc network using queuing model. Int. J. Comput. Technol. Appl. (IJCTA) 2(4) (2011)
7. Narasimhan, B., Santhosh baboo, S.: A hop-by-hop congestion-aware routing protocol for heterogeneous mobile ad-hoc networks. Int. J. Comput. Sci. Inf. Secur. (2009)
8. Hong, C.S., Le, T.A., Abdur Razzaque, Md., et al.: An energy-aware congestion control algorithm for multipath TCP. IEEE Commun. Lett. 16 (2012)
9. Chen, K., Nahrstedt, K., Vaidya, N.: The utility of explicit rate-based flow control in mobile ad hoc networks. In: Proceeding of IEEE, Wireless Communications and Networking Conference (2004)
10. Hasan, S.F., Rahman, K.C.: Explicit rate-based congestion control for multimedia streaming over mobile ad hoc networks. Int. J. Electr. Comput. Sci. 10 (2010)
11. Fang, Y., Zhai, H., Chen, X.: Rate-based transport control for mobile ad hoc networks. In: IEEE Wireless Communications and Networking Conference, vol. 4 (2005)
12. Wang, J, Wen, J., et. al.: An improved TCP congestion control algorithm and its performance. IEEE (2011)
13. Indirani, G., Selvakumar, K.: A swarm-based efficient distributed intrusion detection system for Mobile ad hoc networks. Int. J. Parallel Emerg. Distrib. Syst. (2014)
14. Singh, S.K., Singh, M.P., Singh, D.K.: Routing protocols in wireless sensor networks—a survey. Int. J. Comput. Sci. Eng. Surv. 1 (2010)
15. Jangir, S.K., Hemrajani, N.: A comprehensive review on detection of wormhole attack in MANET. In: International Conference on ICT in Business Industry & Government, Indore (2016)
16. Singh, R.K., Nand, P.: Literature review of routing attacks in MANET. In: International Conference on Computing, Communication and Automation, 2016, pp. 525–530 (2016)

17. Wu, B., Wu, J., Chen, J., Cardei, M.: A survey on attacks and counter measures in mobile ad hoc networks. Springer, Wireless Network Security (2007)
18. Chaurasia, B.K., Tomar, G.S., Shrivastava, L., Bhadoria, S.S.: Secure congestion adaptive routing using group signature scheme. Springer Transaction on Computer Science, vol. 17, 2013, pp. 101–115 (2013)
19. Trivedi, A., Sharma, P.: An approach to defend against wormhole attack in ad hoc network using digital signature. IEEE, pp. 307–311 (2011)
20. Tomar, G.S., Shrivastava, L., Bhadauria, S.S.: Load balanced congestion adaptive routing for randomly distributed mobile ad hoc networks. Springer Int. J. Wirel. Pers. Commun. 75 (2014)

# DDoS Attack Mitigation Using Random and Flow-Based Scheme

Bansidhar Joshi, Bineet Joshi and Kritika Rani

## 1 Introduction

Distributed denial-of-service attacks are a critical threat to the internet. DDoS attackers generate a huge amount of requests to victims through compromised computers (zombies), with the aim of denying normal service or degrading of the quality of services [1]. Internet was primarily designed to facilitate communication and was not designed in a way to give security to such communications. Hence, the network has a lot of scope of being attacked. DoS attacks are growing at a rapid speed, and they are becoming distributed and highly sophisticated. The astounding fact to know is that a DDoS attack can be launched at a minimum of $20. With the help of botnet, DDoS attack can be implemented by sending a few packets by each compromised system. It, hence, becomes difficult to differentiate between legitimate and illegitimate traffic. There are many schemes to tackle these attacks like DPM, PPM, FDPM, and information theory. Several detection schemes exist. Like a technique which prioritizes packets based on a score which estimates its legitimacy given the attribute values it carries and based on this score, selective packet discarding is carried out [2].

B. Joshi (✉) · K. Rani
Jaypee Institute of Information Technology, Noida, India
e-mail: bansidhar.joshi@jiit.ac.in

K. Rani
e-mail: kritika.rani17@gmail.com

B. Joshi
Swami Rama Himalayan University, Dehradun, India
e-mail: bineetjoshi@gmail.com

© Springer Nature Singapore Pte Ltd. 2019
R. K. Shukla et al. (eds.), *Data, Engineering and Applications*,
https://doi.org/10.1007/978-981-13-6351-1_11

Another method was the one in which traffic is analyzed only at the edge routers of an Internet service provider (ISP) network [3, 4]. This framework is able to detect any source-address-spoofed DDoS attack, no matter whether it is a low-volume attack or a high-volume attack. We will be discussing about information theory and linear packet marking technique in this paper and its advantages over conventional DPM and PPM. We will also talk about some of the techniques from which we can prevent our system to be a part of the botnet. Entropy can measure the variations of randomness of flows on a given local router [5]. Based on such variations, a DDoS attack can be detected. If the difference between the entropy and mean is greater than a threshold value, then there is an attack.

## 2 Related Work

### 2.1 Node Append

This is the simplest method of marking to trace the attacker. In this method, we continue appending the information about the router in the header of the packet. The disadvantage is the overhead on the packet header and insufficient space to accommodate so many routers [5].

### 2.2 Node Sampling

This was introduced to solve the problem of storage overhead. The packets are marked depending on the probability chosen at random [5]. Either 0 or 1, can be chosen, so the probability is 0.5. A node once marked is not marked again. The problem lies when it becomes difficult to know which node will mark the packet and a high number of packets are required to trace back the attack.

### 2.3 Edge Sampling

Authors sent the packet with some probability p and overcome the shortcoming in node sampling about the distance from the destination by introducing d. When the packet is marked, it is updated to 0 and is incrementally increased with every decrease in ttl [5]. Once a packet is marked, it is not marked again. $p(1 - p) \wedge (d - 1)$ where d is the number of hops away from the destination.

Let x be the number of packets then

$$E(X) = ln(d)/p(1 - p) \wedge (d - 1) \tag{1}$$

The disadvantage of this technique is that we need many packets to trace the route and in case of multiple attackers, this technique is not that robust.

## 2.4 Probabilistic Packet Marking

This is basically the extension of edge sampling. There is not enough space in the packet header to store the 32-bit address of the router. The value of the router in the fields is fragmented so that it is less used by the router. It can even be applied after or during the attack with no additional router storage problems [6] as in case of logging.

## 2.5 Deterministic Packet Marking

This is almost similar to PPM but as in PPM, the authors mark the packets based on a probability, and in deterministic packet marking, we mark all the packets passing through the network [2]. Using IP address of the router, packets are marked. The problem occurs in case of IP spoofing, where the attackers can spoof their IP address making it difficult to reach the correct router.

## 2.6 Linear Packet Marking

This scheme needs the number of packets almost equal to the hops traversed by the packet [3]. In this trace back scheme, the ID field value is divided by the number of hops. The remainder, thus, obtained is the value of the hop that will mark the packets. Like all packet marking schemes, it also needs an attack to be alive and trace backing to the source itself creates a DoS while performing the attack. In this scheme, the authors use TTL value to decide the hop count and the value in the ID field at the router to decide whether to mark the packet or not.

## 2.7   Entropy Variation

Entropy can measure the variations of randomness of flows on a given local router [7]. Based on such variations, a DDoS attack can be detected. In case of an attack, the actual source can be found by submitting the requests to upstream routers and finally reaching the source. The advantage lies in the fact that it does not require the attack to be alive, can differentiate between legitimate and illegitimate users efficiently, and does not require any marking. The problem in this technique comes in case of trace back when a request is made to all upstream routers in order to identify the source.

## 3   Analysis of a Random and Flow-Based Model

We found out that the maximum hops that a packet travels before reaching the destination is 25. A simulation is done using the tracert command and the same is shown in Figs. 1 and 2. The algorithm that we are using for detecting whether a DoS or a DDoS is happening or is it a surge in the legitimate traffic is explained in Algorithm 1, and the algorithm we are using for marking the packet for successful IP trace back is explained in Algorithm 2.

```
C:\Users\sss>tracert www.facebook.com

Tracing route to www.facebook.com [31.13.78.35]
over a maximum of 30 hops:

  1      2 ms      1 ms      1 ms   192.168.1.1
  2     17 ms     17 ms     17 ms   abts-north-static-073.220.160.122.airtelbroadban
d.in [122.160.220.73]
  3      *         *         *      Request timed out.
  4     17 ms     17 ms     16 ms   abts-north-static-085.176.144.59.airtelbroadband
.in [59.144.176.85]
  5    124 ms    126 ms    139 ms   182.79.255.106
  6    110 ms     93 ms     95 ms   182.79.234.110
  7     87 ms     81 ms     89 ms   182.79.236.114
  8    102 ms     95 ms     96 ms   182.79.247.194
  9     94 ms     94 ms     94 ms   182.79.222.106
 10     95 ms     94 ms     95 ms   ae17.pr01.sin1.tfbnw.net [103.4.97.138]
 11     95 ms     95 ms    109 ms   ae11.bb01.sin1.tfbnw.net [31.13.27.152]
 12    155 ms     98 ms     96 ms   ae5.bb01.sgp1.tfbnw.net [31.13.24.124]
 13     98 ms     99 ms     98 ms   ae1.ar01.sin4.tfbnw.net [74.119.76.43]
 14     98 ms     99 ms    119 ms   psw01d.sin4.tfbnw.net [173.252.65.98]
 15    105 ms     95 ms     96 ms   msw1af.01.sin4.tfbnw.net [204.15.21.0]
 16     96 ms     99 ms     98 ms   edge-star-mini-shv-01-sin4.facebook.com [31.13.7
8.35]

Trace complete.
```

**Fig. 1** Tracing the route of Facebook.com

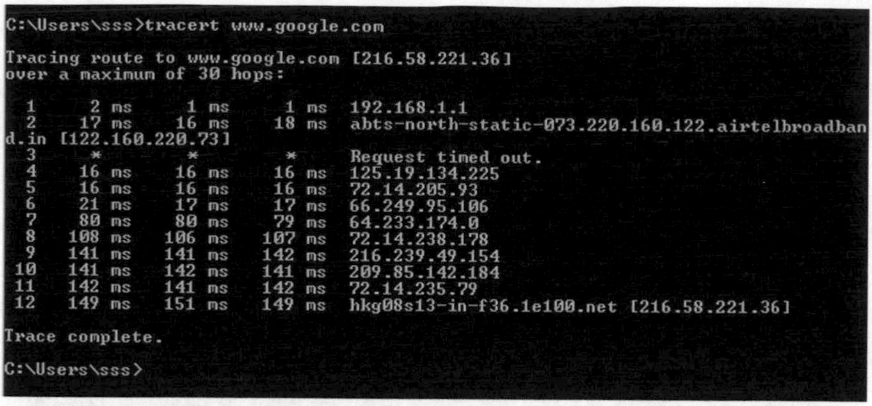

```
C:\Users\sss>tracert www.google.com

Tracing route to www.google.com [216.58.221.36]
over a maximum of 30 hops:

  1     2 ms     1 ms     1 ms  192.168.1.1
  2    17 ms    16 ms    18 ms  abts-north-static-073.220.160.122.airtelbroadban
d.in [122.160.220.73]
  3     *        *        *     Request timed out.
  4    16 ms    16 ms    16 ms  125.19.134.225
  5    16 ms    16 ms    16 ms  72.14.205.93
  6    21 ms    17 ms    17 ms  66.249.95.106
  7    80 ms    80 ms    79 ms  64.233.174.0
  8   108 ms   106 ms   107 ms  72.14.238.178
  9   141 ms   141 ms   142 ms  216.239.49.154
 10   141 ms   142 ms   141 ms  209.85.142.184
 11   142 ms   141 ms   142 ms  72.14.235.79
 12   149 ms   151 ms   149 ms  hkg08s13-in-f36.1e100.net [216.58.221.36]

Trace complete.

C:\Users\sss>
```

**Fig. 2** Tracing the route of Google

## 3.1 Algorithm Used for This Work Are

**Algorithm 1**: To determine the threshold time after which we monitor if there is an attack.

Start from 1 time unit after which we will check for an attack and increase the time exponentially. The next time interval after which we monitor the attack will be 2 units. So the time intervals are increasing like this 1, 2, 4...

A threshold value is formed based on past experiment and trends keeping in mind (based on different kinds of networks for different purposes like in an airline website, No. of visits during festivals or holidays increases to many folds):

Different months
Days
Seasons (summer, winter, spring, rainy, etc.) Festivals (can be included in months)
Weekly

On reaching this threshold value, we start increasing the time interval additively instead of exponentially, i.e., now, time interval increases by 1 time unit...

$$k, k + 1, k + 2 \ldots$$

If there as an attack at any point, mitigate it, reduce the threshold value to half of the time interval, and reduce the time to 1 unit again and repeat the process.

**Algorithm 2**: To detect an attack due to high number of packets being sent by a few attackers, local router is the router taken into consideration. Upstream routers are the routers just above the local router. A flow is defined as fij <ui, dj, t>: ui is an upstream router, dj is the destination address, and t is the current time.

The algorithm is as follows [8]:

1. Initialize C0, δ0 with some experimental value.
2. Identify the flows f1, f2, f3 … fn and set count number of each flow to zero, x1 = 0, x2 = 0, x3 = 0 …, xn = 0 (count number is basically the number of times a particular flow (having same upstream router and timestamp) occurs in a given time).
3. When time interval used to monitor the traffic (as described in the above algorithm) is over, calculate the probability distribution and entropy variation as follows:

$$pj = xj * (\sum_{j=1}^{n} x) - 1 \tag{2}$$

$$h = -\sum_{j=1}^{n} pj \log(pj) \tag{3}$$

4. Save x1, x2 … xn and H(F).
5. If there is no dramatic change of the entropy variation H(F), |H(F)-C| <= δ update mean and the standard variation is

$$C[t] = \sum_{j=1}^{n} \alpha j * C[t-j], \qquad \sum_{j=1}^{n} \alpha j = 1, \tag{4}$$

$$\delta[t] = \sum_{j=1}^{n} \beta j * \delta[t-j], \qquad \sum_{j=1}^{n} \beta j = 1, \tag{5}$$

6. Go to step 2.
7. If |H(F)-C| > δ, then the attack is possible.
8. Now if pi * log(pi/qj) > 0.2 (where qj is another flow), then it is a legitimate traffic. Otherwise, it is an attack.

| Versi | H. | TOS(8) | Total length(16) | | |
|-------|-----|--------|-------|---------|------|
| Identification(16) | | | Flags | Fragmentation Offset(12) | |
| Time to Live(8) | | Protocol (8) | Header Checksum(8) | | |
| | | | | | |
| 32-bit destination address | | | | | |
| | | | | | |

**Fig. 3** IPv4 header

**Algorithm 3**: We are implementing the algorithm based on the assumption that a packet on an average takes only 25 hops to reach the destination [6]. The steps of marking are as follows: on the arrival of a packet, we will apply our packet marking algorithm. Then, based on the algorithm, we will see to mark the packet or leave it unmarked and finally forward it on the network [2]. We use ID field of the ipv4 to store the unique identification number of the router, as well as the maximum hop distance from the destination. A diagram of the complete header is given in Fig. 3. Distance is calculated using the ttl value of the packet that tells the maximum time the packet can stay on the network.

1. If packet is not marked
2.     Select random number'x' between 1 and 25
3.     Dist away= Dist – ttl
4.     If((packet ID % x) + 1 == identification number of a router)
5.         Insert router address into packet header
6.         Set MRN = 0
7.     Endif
8. Else
9.     Increment MRN
10. End if

We simulated our detection Algorithm 1 using Wireshark. We edited our files by converting them into text format using editcap, modifying the header using tcprewrite, used multithreading to make the capturing and detection modules run together, and tcpreplay to resend the packets. Combining all such concepts in C programming on a Linux platform, we were able to implement our algorithm. The calculations are shown below. The detailed algorithm is explained in Fig. 4.

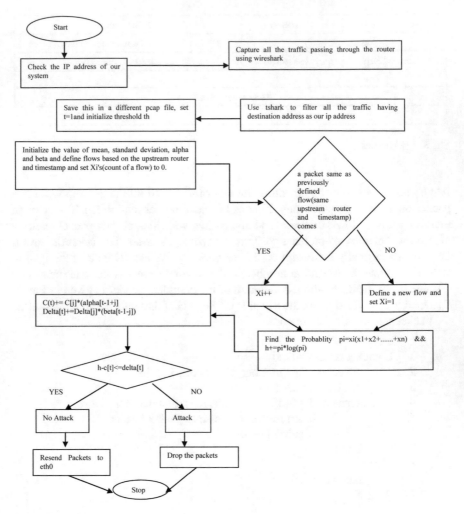

**Fig. 4** Flow diagram

The simulation of Algorithm 2 is shown using ns2. The changes are made in the header of the trace file. The data needed is saved in .tr file and is extracted for analyzing using the awk script on the desired columns.

**Case 1**

X1 = 5        X2 = 4   X3 = 6   X4 = 4

P1 = 5/21         P2 = 6/21              P3 = 6/21

           P4 = 4/21

H+=Pi * log(Pi)

P1 * log(P1) = -0.148    P2*log(P2) = -0.155

P3*log(P3) = -0.155      P4*log(P4) = -0.137

H = -0.595      c[0] = 0.2      delta[0] = -0.4

H - c[t] <= delta[t]

H – c[0] <= delta[0]

-0.795 <= -0.4   true

Therefore, no attack.

**Case 2**

X1 = 100       X2 = 4          X3 = 2

P1 = 0.943      P2 = 0.037    P3 = 0.018

H+=Pi * log(Pi)

H = -0.1083     c[0] = 0.2      delta[0] = -0.4

H - c[t] <= delta[t]

H – c[0] <= delta[0]

-0.308 <= -0.4   false

Therefore, attack.

# 4   Conclusion and Future Work

## 4.1   Conclusion

The conclusions drawn from the paper are that DoS and DDoS attack are a great threat to the internet. Packet marking techniques are costly to implement and are difficult to predict at an early stage. Moreover, we need the attack to be alive to trace back to the origin of the attack. The entropy variation technique is thus a relatively new technique for the detection of DDoS and DoS attack and there is a steep learning curve for the whole organization, for both users and administrators. Packet marking techniques, on the other hand, are more mature, well documented, and easy to understand.

For the people who are not familiar with the technology used for information theory should adopt to packet marking techniques for the reason stated above and wait for the information theory to become more mature.

## 4.2  Future Work

In future, we would like to implement these techniques on the public and private cloud setup and compare the results of our algorithm with the results of the existing techniques. We would also check the scalability and the fault tolerance of the algorithm in both the types of cloud that is public as well as private cloud.

# References

1. Sagar. A., Joshi B. K., Mathur, N.: A study of distributed denial of service attack in cloud computing (DDoS). In: Edition on Cloud and Distributed Computing: Advances and Applications, vol. 2 (2013)
2. Belenky, A., Ansari, N.: IP traceback with deterministic packet marking. Commun. Lett. IEEE **7**(4), 162–164 (2003)
3. Saurabh, S., Sairam, A.S.: Linear and Remainder: Packet Marking for Fast IP TraceBack. IEEE. 978-1-4673-0298-2/12/ (2012)
4. Joshi, B., Joshi, B., Rani, K.: Mitigating data segregation and privacy issues in cloud computing. In: Proceedings of International Conference on Communication and Networks: ComNet 2016, vol. 508, pp. 175. Springer (2017)
5. Savage, S., Wetherall, D., Karlin, A., Anderson, T.: Network support for IP traceback. IEEE Trans. Netw. **9**(3) (2001)
6. Goodrich, M.T.: Probabilistic packet marking for large-scale IP traceback. IEEE/ACM Trans. Netw. **16**(1), 15–24 (2008)
7. Yu, S., Zhou, W., Doss, R.: Information theory based detection against network. behavior mimicking DDoS attacks. IEEE Commun. Lett. **12**(4) (2008)
8. David, J., Thomas, C.: DDoS attack detection using fast entropy approach on flow-based network traffic. Procedia Comput. Sci. **5**, 30–36 (2015)

# Digital Image Watermarking Against Geometrical Attack

Sandeep Rai, Rajesh Boghey, Dipesh Shahane and Priyanka Saxena

## 1 Introduction

The advancement of varied networks led to the movement to packet info. On these lines, transmission parts get to be clearly accessible to very large range of users. Purchasers will get to information and, besides, management them in several ways in which. Various ways for golf shot away and replicating info will disregard the copyrights. Contingent upon the weather, any authors will decide two types of watermarks. Digital watermarking could be a system of infusion the digital mark whereas exchanging of sure parts. The most reasonable watermark is associate degree clear watermark. Noticcable watermark is plainly obvious to any shopper of the substance. The resistance of those watermarks is questionable. Manufacturing a large range of duplicates of a selected substance will undermine or maybe obliterate the character of the watermark. The second method is embedding the watermark so once its consolidation it gets to be clearly impalpable. This technique is distinctive and additionally baffled than the past one. The impalpable watermark is additionally dependable than perceptible watermark.

Watermarking infers the existence of a mark in transmission substance that contains the name, his signature or mark. Client of substance cannot see the inserted watermark [1]. Algorithmic rule for embedding associate degree impalpable

S. Rai · R. Boghey · D. Shahane (✉) · P. Saxena
Computer Science and Engineering, Technocrats Institute Technology (Excellence),
Bhopal, India
e-mail: dipeshshahane@gmail.com

S. Rai
e-mail: sandtec@gmail.com

R. Boghey
e-mail: rajeshboghey@gmail.com

P. Saxena
e-mail: priyanka_0789@yahoo.com

© Springer Nature Singapore Pte Ltd. 2019
R. K. Shukla et al. (eds.), *Data, Engineering and Applications*,
https://doi.org/10.1007/978-981-13-6351-1_12

watermark depends upon the engravings of the watermark within the frequency domain. Within the frequency domain watermarking is more durable to separate while not abusing the character of the watched image.

The application of the three basic types of transformation: Fast Fourier Transform (FFT), Discrete Cosine Transform (DCT), and Discrete wavelet Transform (DWT) square measure displayed within the literature [2–4]. It has been incontestable that DWT acquires the foremost current outcomes as well as associate degree undetectable watermark. Despite the strategy for info appropriation and also the amount of duplicates created, the chief objective of the watermark engraving is to accomplish validation.

The author has introduced various ways to watermark the photographs. This paper is focused on watermark embedding supported DT-CWT and cryptography technique. Contingent upon the connected unfold spectrum ways sure algorithmic rule can have comparable qualities as once considering a radio framework that applies the vital procedures of unfold spectrum. The final method for watermarking is given by Fig. 1 (Figs. 2 and 3).

The paper is formulated star as follows. In Sect. 1, the introduction concerning watermarking and unfold spectrum is mentioned. Digital watermarking methods

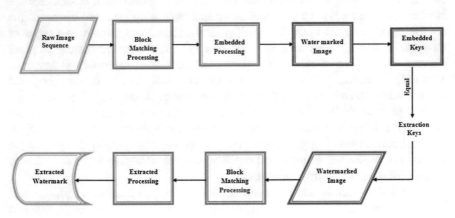

**Fig. 1** Process of watermarking

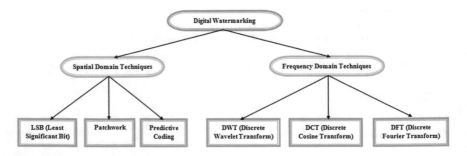

**Fig. 2** Classification of digital watermarking

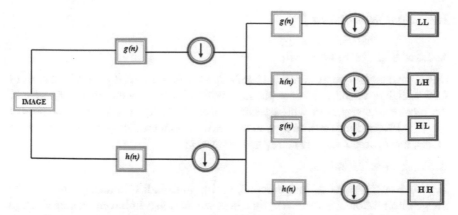

**Fig. 3** 2D-DWT decomposition of an input image using filtering approach

are discussed in Sect. 2. The planned methodology and its connected algorithmic rule square measure best owed in Sect. 3. In Sect. 4, experimental and its analysis is between performance metrics is discussing and last remarks square measure bestowed in Sect. 4.

## 2 Watermarking Techniques

There square measure several algorithms that square measure being employed to cover the key information. These algorithms are categorized into two domains, which are given as follows:

1. Spatial domain and
2. Frequency domain

Spatial domain watermarking slightly modifies the pixels of 1 or 2 indiscriminately elect subsets of a picture. On the opposite aspect, in frequency domain techniques the image is first remodeled to the frequency domain by the utilization of any demodulation ways like Fourier transform, distinct circular function remodel (DCT) or distinct rippling remodel (DWT). Currently, the knowledge is another to the values of its remodel coefficients. Once applying the inverse remodel, the marked coefficients kind the embedded image.

## 2.1 Spatial Domain [5]

### A. Least Significant bit (LSB)

During this technique, watermark is embedded within the LSB of pixels. 2 sorts of LSB techniques square measure planned. Within the first methodology, the LSB of the picture was recouped with a pseudo-noise (PN) sequence whereas within the second a PN sequence was another to the LSB. This methodology is straightforward to use, however, not terribly sturdy against attacks.

### B. Patchwork Technique

In patchwork, n pairs of image points, (a, b), were indiscriminately chosen. The image information in a very were lightened whereas that in b were darkened. High level of hardiness against many sorts of attacks square measure provided during this technique. However, here during this technique, terribly bit of data is hidden.

### C. Predictive Coding Scheme

In this scheme, a pseudorandom noise (PN) pattern says $W(x, y)$ is another to hide image. It will increase the hardiness of watermark by increasing the gain issue. However, as a result of high increment in gain issue, image quality could decrease.

### 2.1.1 Frequency Domain

### A. Discrete Cosine Transform [6]

The high-frequency components are watermarked in frequency domain. The main steps are given as follows:

(1) Divide the image into non-overlapping blocks of $8 \times 8$
(2) Apply forward DCT to each of these blocks
(3) Apply some block selection criteria (e.g., HVS)
(4) Apply coefficient selection criteria (e.g., highest)
(5) Embed watermark by modifying the selected coefficients.
(6) Apply inverse DCT transform on each block.

### B. DFT Domain Watermarking

DFT Domain Watermarking DFT domain is the favorite alternative of researches as a result of it provides hardiness against geometric attacks like translation, rotation, cropping, scaling, etc. There square measure 2 sorts of DFT based mostly watermark embedding techniques. In first technique watermark is directly inserted and the other technique is an example based mostly embedding. In direct embedding, watermark is inserted by ever-changing the section info at intervals the DFT [7].

An example could be a structure that is employed within the DFT domain to evaluate the transformation issue. First, a change is created in image then to resynchronize the image this example is searched, then use the detector to extract the embedded unfold spectrum watermark.

## C. Discrete Wavelet Transform

Discrete wavelet transform is applied to decompose any non-stationary signal like a picture, audio or video signal. The remodel is based on very little waves, referred to as wavelets of varied frequency and restricted period. Frequency also as spatial info of a picture is maintained throughout rippling transformation. Temporal info is preserved throughout this conversion methodology [8]. Wavelets square measure created by transformations and dilations of constant perform referred to as mother rippling. DWT is accomplished by low-pass and high-pass filtering of a picture. High-pass filter creates elaborated image pixels and low-pass filter creates coarse approximation image pixels [9]. The outputs square measure down-sampled by two once acting the low-pass and high-pass filtering. Second DWT is finished by death penalty 1DDWT on every row that is thought as horizontal filtering then on every column that is thought as vertical filtering [10].

# 3 Literature Review

In previous few years, a many watermarking techniques are evolved within the history of watermarking. The researchers have examined the algorithms on distinct parameters like capability, strength physical property, etc., a number of the examinations determined by the varied researchers includes associate degree algorithmic rule that uses each digital image watermarking and digital signature to produce integrity which will be verified by user at the network and that they have instructed a changed algorithmic rule that aims that within the mixed hybrid transformation once the quilt image is altered within the singular values instead of on DWT subbands, so it makes the watermark image additionally unsafe toward various attacks, whereas PSNR of each the image are increased. [11] at that time another researches additionally propounded a way that increased verification for integrity of information over the network. [12] Next few researches mentioned concerning the combined DWT DCT transformation with low-frequency digital watermarking. The experimental outcome holds the potential to tolerate geometric attacks and customary signal process. [13] at that time researches planned a strong protection technique that was smitten by (DWT) and visual hided theme (VHI). The outcomes were tested on parameters PSNR and resistance against completely different attacks [14]. Additionally, at that time some researchers instructed a way in spatial domain watermarking as well as secret writing techniques. The results were judged on the idea of the standard of original image and square measure thought of to be satisfactory. [15] additional enhancements and changes is created in existing algorithms for higher performance in each field, however, still expecting that square measure algorithms will provide higher leads to each field appears extremely tough.

## 4  Proposed Methodology

This section provides temporary summary of the planned methodology DT-CWT is discussing. The Dual-Tree complicated rippling remodel (DT-CWT) has been introduced to beat the disadvantages of real DWT. The overall execution of the DT-CWT style ensures the subsequent properties:

- Approximate shift invariance,
- Good directional selectivity in 2D with Gabor-like filters also true for higher dimensionality (m-D),
- Perfect reconstruction using short linear-phase filters,
- Limited redundancy: independent of the number of scales: 2:1 for 1-D (2 m:1) for m-D,
- Efficient order-N computation - only twice the simple DWT for 1-D (2 m times for m-D);

The overall performance of the DT-CWT design ensures the shift invariance property of the transformation. Moreover, it improves the directional selectivity compared to the DWT since it produces six directional subbands at each scale oriented at $\pm 15°$, $\pm 45°$, $\pm 75°$ compared to the three directional subbands of the DWT. Figure 4 shows a two-level decomposition of 1-D signal f(x) using DT-CWT (Figs. 5 and 6).

A DT-CWT transformation of 1D signal f(x) in terms of shifted and dilated wavelet function $\varphi(n)$ and scaling function $\emptyset(n)$ is given by the following equation:

$$f(x) = \sum_{l \in Z} S_{j_0,l} \phi_{j_0,l}(x) + \sum_{j \ge j_0} \sum_{l \in Z} c_{j,l} \psi_{j,l}(x) \tag{1}$$

where Z is the set of natural numbers, j and l refer to the index of shifts and dilations, respectively, is the scaling coefficient, and is the complex wavelet coefficient with

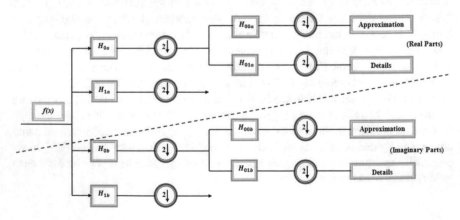

**Fig. 4**  DT-CWT

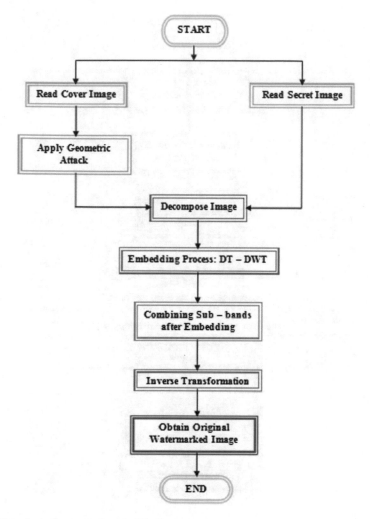

**Fig. 5** Flowchart of watermark embedding

$$\phi_{j_{0,l}}(x) = \phi^r_{j_{0,l}}(x) + \sqrt{-1}\phi^i_{j_{0,l}}(x) \, and \, \psi_{j_{0,l}}(x) = \psi^r_{j_{0,l}}(x)\sqrt{-1}\psi^i_{j_{0,l}}(x)$$

where the superscripts r and i denote the real and imaginary parts, respectively. To compute the 2-D DT-CWT of images, the two trees are applied to the rows and then to the columns of the image as in the basic DWT. This operation results in six complex high-pass subbands at each level and two complex low-pass subbands on which subsequent stages iterate. The decomposition of 2D signal can be expressed in the same manner like the 1D decomposition in [7] as follows:

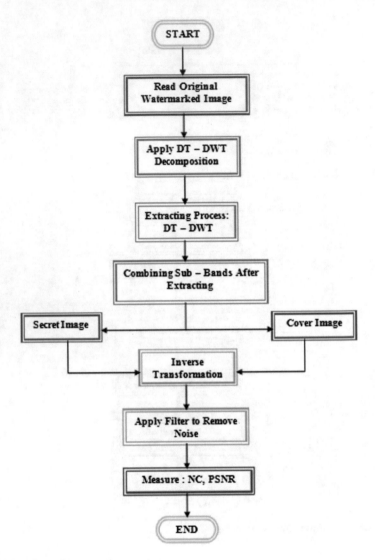

**Fig. 6** Flowchart of watermark extraction

$$f(x) = \sum\nolimits_{l \in Z^2} S_{j_0,l} \phi_{j_0,l}(x, y) + \sum\nolimits_{\theta \in \theta} \sum\nolimits_{j \geq j_0} \sum\nolimits_{l \in Z^2} C_{j,l}^{\theta} \psi_{j,l}^{\theta}(x, y) \qquad (2)$$

where $\theta \in \Theta = \{\pm 15°, \pm 45°, \pm 75°\}$ which determine the complex wavelet directionality.

# 5 Experimental Results and Analysis

The tentative study of the planned methodology is finished employing a widely used MATLAB2012A tool cabinet and also the machine configuration is Intel I3 core two.20 Ghz processor, with 4 GB RAM, windows seven home basis. In planned methodology, we have a tendency to apply a compression and secret writing and watermarking for the surefire bar from geometric attacks.

## 5.1 Snapshots

The figures shown below are that of the original image that must be watermarked, and screenshot of main interface is additionally shown in Fig. 7. The planned feature image is river that is revolved, translated, and sheared concerning the angle of ten, twenty, and thirty degree. The interface of these options is additionally shown through figure (Figs. 8, 9, 10, 11, 12, 13, 14, 15 and 16).

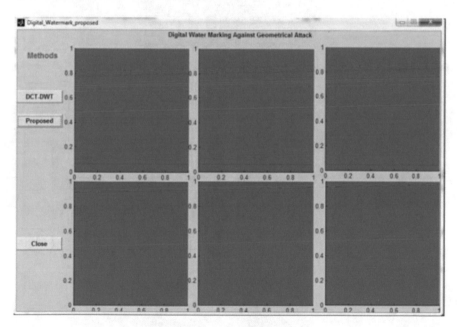

**Fig. 7** Screenshot of main GUI

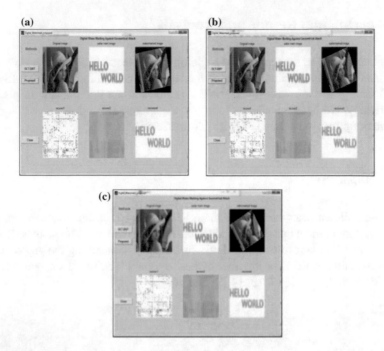

**Fig. 8** Rotation of Leena about 10, 20, and 30°

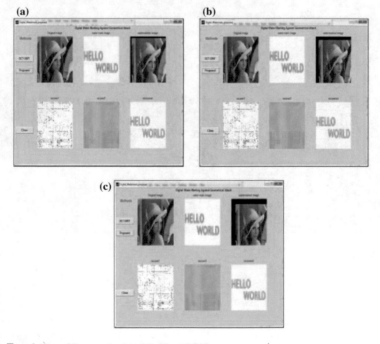

**Fig. 9** Translation of Leena at value 10, 15, and 20°

**(a)** **(b)**

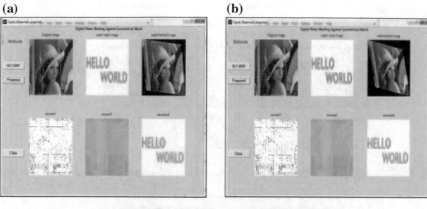

**(c)**

**Fig. 10** Shear of Leena at value 0.15, 0.25, and 0.35

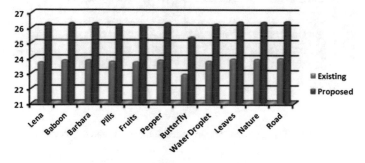

**Fig. 11** Comparison of PSNR values between proposed and existing methodology (at $10^0$ rotation)

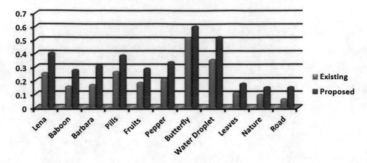

**Fig. 12** Comparison of NC values between proposed and existing methodology (at $10^0$ rotation)

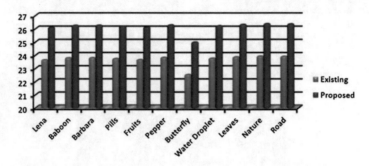

**Fig. 13** Comparison of PSNR values between proposed and existing methodology (at value 10 translation)

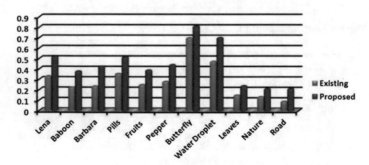

**Fig. 14** Comparison of NC values between proposed and existing methodology (at value 10 translation)

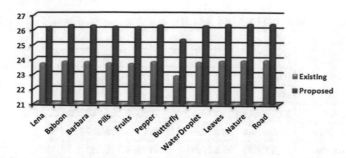

**Fig. 15** Comparison of PSNR values between proposed and existing methodology (at value 10 shear)

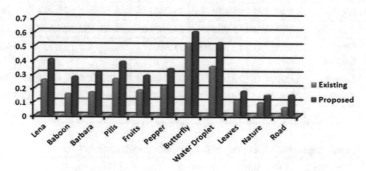

**Fig. 16** Comparison of NC values between proposed and existing methodology (at value 10 Shear)

## 5.2 Result Analysis

The comparative study of the planned methodology is perform victimization the transformation metrics rotation, shear, and translation and also the simulation results of it shown in Tables 1, 2, 3, 4, 5, and 6.

## 6 Conclusion

In scientific method, during this analysis we have a tendency to gift a strong image watermarking algorithmic rule against numerous geometric attacks mathematically invariant to the cutting, translation, and rotation concerning ten, twenty, and thirty degree. A hybrid DT-CWT and cryptography technique for digital watermarking system wavelets was bestowed that incontestable sensible performance underneath numerous geometric attacks. The experimental analysis is engaged on river, Baboon, Barbara, Pills, fruits, Pepper, Butterfly, Water driblet, Leaves, Nature, and Road feature pictures. This methodology proves to be a much better technique leading to the many improvement in PSNR and Old North State activity parameter.

**Table 1** Result analysis for PSNR of proposed and existing method (rotation)

| S. No. | Images | Existing | | | Proposed | | |
|--------|--------|----------|--|--|----------|--|--|
| | | PSVR | | | PSXR | | |
| | | Angle | | | Angle | | |
| | | 10° | 20° | 30° | 10° | 20° | 30° |
| 1 | Lena | 23.690442 | 23.723472 | 23.758009 | 26.287206 | 26.287386 | 26.294237 |
| 2 | Baboon | 23.834364 | 23.834544 | 23.841395 | 26.287206 | 26.287386 | 26.294237 |
| 3 | Barbara | 23.832627 | 23.839941 | 23.849818 | 26.285469 | 26.292783 | 26.302660 |
| 4 | Pills | 23.736836 | 23.750683 | 23.771723 | 26.189678 | 26.203525 | 26.224565 |
| 5 | Fruits | 23.709867 | 23.750487 | 23.783580 | 26.162709 | 26.203329 | 26.236422 |
| 6 | Pepper | 23.836499 | 23.839188 | 23.845114 | 26.289341 | 26.292030 | 26.297956 |
| 7 | Butterfly | 22.903890 | 23.082431 | 23.183102 | 25.356732 | 25.535273 | 25.635944 |
| 8 | Water droplet | 23.765383 | 23.772806 | 23.800175 | 26.218225 | 26.225648 | 26.253017 |
| 9 | Leaves | 23.916408 | 23.935774 | 23.946373 | 26.369250 | 26.388616 | 26.399215 |
| 10 | Nature | 23.897798 | 23.888705 | 23.898559 | 26.350640 | 26.341547 | 26.351401 |
| 11 | Road | 23.951449 | 23.963869 | 23.975304 | 26.404291 | 26.416711 | 26.428146 |

**Table 2** Result analysis for NC of proposed and existing method (rotation)

| S. No. | Images | Existing | | | Proposed | | |
|--------|--------|----------|--|--|----------|--|--|
| | | NC | | | NC | | |
| | | Angie | | | Angle | | |
| | | 10° | 20° | 30° | 10° | 20° | 30° |
| 1 | Lena | 0.255750 | 0.200236 | 0.170715 | 0.282558 | 0.224462 | 0.194986 |
| 2 | Baboon | 0.164596 | 0,130370 | 0.113064 | 0.282558 | 0.224462 | 0.194986 |
| 3 | Barbara | 0.16"451 | 0.134933 | 0.117885 | 0,307476 | 0.247094 | 0.216201 |
| 4 | Pills | 0.259130 | 0.214548 | 0.18220" | 0.379050 | 0.310941 | 0.266400 |
| 5 | Fruits | 0.184037 | 0.147744 | 0.124266 | 0.287316 | 0.231792 | 0.196530 |
| 6 | Pepper | 0.217269 | 0.169753 | 0.145467 | 0.340100 | 0.267992 | 0.230501 |
| 7 | Butterfly | 0.516316 | 0,409299 | 0.351918 | 0,602655 | 0,480336 | 0.414514 |
| 8 | Water droplet | 0.350765 | 0.277955 | 0.241719 | 0.520610 | 0.413987 | 0.359411 |
| 9 | Leaves | 0.108027 | 0.086161 | 0.075077 | 0.173377 | 0.139036 | 0.121060 |
| 10 | Nature | 0.094880 | 0,074933 | 0.064264 | 0.151297 | 0.119007 | 0.102404 |
| 11 | Road | 0.064249 | 0.052124 | 0.044108 | 0.152737 | 0.122227 | 0.104145 |

**Table 3** Result Analysis for PSNR of Proposed and Exiting Method (Translation)

| S. No. | Images | Existing | | | Proposed | | |
|---|---|---|---|---|---|---|---|
| | | PSNR | | | PSNR | | |
| | | Value | | | Value | | |
| | | 10 | 15 | 20 | 10 | 15 | 20 |
| 1 | Lena | 23.643036 | 23.653643 | 23.673332 | 26.095878 | 26.106485 | 26.126174 |
| 2 | Baboon | 23.794942 | 23.90812 | 23.785124 | 26.247784 | 26.243654 | 26.237966 |
| 3 | Barbara | 23.805621 | 23.808931 | 23.813349 | 26.258463 | 26.261773 | 26.266191 |
| 4 | Pills | 23.733215 | 23.734345 | 23.735299 | 26.186057 | 26.187187 | 26.188141 |
| 5 | Fruits | 23.659369 | 23.669687 | 23.678836 | 26.112211 | 26.122529 | 26.131678 |
| 6 | Pepper | 23.831100 | 23.833891 | 23.836433 | 26.283942 | 26.286733 | 26.289275 |
| 7 | Butterfly | 22.524690 | 22.536192 | 22.5377S9 | 24.977532 | 24.989034 | 24.990631 |
| 8 | Water droplet | 23.769773 | 23.773554 | 23.775670 | 26.222615 | 26.226396 | 26.228512 |
| 9 | Leaves | 23.871067 | 23.871386 | 23.872758 | 26.323909 | 26.324228 | 26.325600 |
| 10 | Nature | 23.925209 | 23.930791 | 23.936176 | 26.378051 | 26.383633 | 26.389018 |
| 11 | Road | 23.909259 | 23.909065 | 23.909572 | 26.362101 | 26.361907 | 26.362414 |

**Table 4** Result analysis for NC of proposed and existing method (translation)

| S. No. | Images | Existing | | | Proposed | | |
|---|---|---|---|---|---|---|---|
| | | NC | | | NC | | |
| | | Value | | | Value | | |
| | | 10 | 15 | 20 | 10 | 15 | 20 |
| 1 | Lena | 0.330870 | 0.316915 | 0.302370 | 0.516591 | 0.495148 | 0.473483 |
| 2 | Baboon | 0.223106 | 0.219773 | 0.216638 | 0.377121 | 0.369314 | 0.361971 |
| 3 | Barbara | 0.234663 | 0.232567 | 0.230225 | 0.423452 | 0.417654 | 0.411254 |
| 4 | Pills | 0.354529 | 0.348840 | 0.342604 | 0.513315 | 0.504545 | 0.494942 |
| 5 | Fruits | 0.248633 | 0.244956 | 0.240612 | 0.386625 | 0.380715 | 0.373929 |
| 6 | Pepper | 0.279385 | 0.269843 | 0.261077 | 0.439573 | 0.427250 | 0.415598 |
| 7 | Butterfly | 0.698425 | 0.692499 | 0.687426 | 0.810145 | 0.802810 | 0.796203 |
| 8 | Water droplet | 0.471893 | 0.464648 | 0.458068 | 0.698577 | 0.688885 | 0.679935 |
| 9 | Leaves | 0.149112 | 0.147514 | 0.146070 | 0.237602 | 0.235195 | 0.233028 |
| 10 | Nature | 0.133566 | 0.133342 | 0.132953 | 0.209118 | 0.207878 | 0.206496 |
| 11 | Road | 0.089387 | 0.089150 | 0.088802 | 0.210013 | 0.208578 | 0.207093 |

**Table 5** Result analysis for PSNR of proposed and existing method (shear)

| S. No. | Images | Existing | | | Proposed | | |
|--------|--------|----------|---|---|----------|---|---|
| | | PSNR | | | PSNR | | |
| | | Value | | | Value | | |
| | | 0.15 | 0. 25 | 0.35 | 0.15 | 0.25 | 0.35 |
| 1 | Lena | 23.715015 | 23.752814 | 23.790334 | 26.167857 | 26.205656 | 26.243176 |
| 2 | Baboon | 23.850482 | 23.867665 | 23.879131 | 26.303324 | 26.320507 | 26.331973 |
| 3 | Barbara | 23.822617 | 23.845693 | 23.873258 | 26.275459 | 26.298535 | 26.326100 |
| 4 | Pills | 23.742306 | 23.765553 | 23.804496 | 26.195148 | 26.218395 | 26.257338 |
| 5 | Fruits | 23.715199 | 23.760868 | 23.796230 | 26.168041 | 26.213710 | 26.249072 |
| 6 | Pepper | 23.842548 | 23.856070 | 23.865659 | 26.295390 | 26.308912 | 26.318501 |
| 7 | Butterfly | 22.891702 | 23.084908 | 23.262813 | 25.344544 | 25.537750 | 25.715655 |
| 8 | Water droplet | 23.804681 | 23.832581 | 23.851985 | 26.257523 | 26.285423 | 26.304827 |
| 9 | Leaves | 23.897107 | 23.920286 | 23.949555 | 26.349949 | 26.373128 | 26.402397 |
| 10 | Nature | 23.921532 | 23.927855 | 23.937000 | 26.374374 | 26.380697 | 26.389842 |
| 11 | Road | 23.933676 | 23.939348 | 23.948645 | 26.386518 | 26.392190 | 26.401487 |

**Table 6** Result analysis for NC of proposed and existing method (shear)

| S. No. | Images | Existing | | | Proposed | | |
|--------|--------|----------|---|---|----------|---|---|
| | | NC | | | NC | | |
| | | Value | | | Value | | |
| | | 0.15 | 0.25 | 0.35 | 0.15 | 0.25 | 0.35 |
| 1 | Lena | 0.254491 | 0.203415 | 0.157541 | 0.401799 | 0.312431 | 0.252571 |
| 2 | Baboon | 0.153802 | 0.121434 | 0.095522 | 0.276414 | 0.219083 | 0.173202 |
| 3 | Barbara | 0.16(5150 | 0.1347(56 | 0.106613 | 0.3097'3 | 0.250031 | 0.198201 |
| 4 | Pills | 0.262876 | 0.213267 | 0.166833 | 0.384710 | 0.311263 | 0.244572 |
| 5 | Fruits | 0.181774 | 0.145586 | 0.114336 | 0.286692 | 0.230623 | 0.182667 |
| 6 | Pepper | 0.211567 | 0.1(5644(5 | 0.119818 | 0.334872 | 0.265092 | 0.207252 |
| 7 | Butterfly | 0.515929 | 0.412427 | 0.322568 | 0.600815 | 0.482222 | 0.379185 |
| 8 | Water droplet | 0.353313 | 0.231437 | 0.219157 | 0.521636 | 0.416736 | 0.326777 |
| 9 | Leaves | 0.110082 | 0.087076 | 0.066542 | 0.175870 | 0.139792 | 0.107647 |
| 10 | Nature | 0.093815 | 0.074305 | 0.053466 | 0.143694 | 0.118264 | 0.003006 |
| 11 | Road | 0.060782 | 0.040130 | 0.038622 | 0.149112 | 0.119493 | 0.093769 |

# References

1. Samcovic, A., Turan, J.: Attacks on digital wavelet image watermarks. J. Electr. Eng. **59**(3), 131–138 (2008)
2. Shi, Y.Q., Sun, H.: Image and Video Compression for Multimedia Engineering: Fundamentals, Algorithms and Standards. Taylor & Francis Group, Boca Raton (2008)
3. Hameed, K., Mumtaz, A., Gilani, S.A.M.: Digital image watermarking in the wavelet transform domain. World Acad. Sci. Eng. Technol. **13**, 86–89 (2006)
4. Chawla, K., Singh, S.: Comparative analysis of watermarking techniques using frequency domain and wavelet domain technologies. Int. J. Comput. Eng. Manag. **15**(5), 74–76 (2012)
5. Shukla, D., Tiwari, N., Dubey, D.: Survey on digital watermarking techniques. Int. J. Signal Process. Image Process. Pattern Recognit. **9**(1), 239–244 (2016)
6. Kushwah, V.R.S., Tiwari, S., Gautam, M.: A review study on digital watermarking techniques. Int. J. Curr. Eng. Sci. Res. **3**(1) (2016). ISSN (Print): 2393-8374, (Online): 2394-0697
7. Potdar, V.M., Han, S., Chang, E.: A survey of digital image watermarking techniques. In: 2005 3rd IEEE International Conference on Industrial Informatics, INDIN'05, 10–12 Aug 2005, pp.709–716 (2005)
8. Kaur, R., Jindal, S.: Robust digital watermarking in high frequency band using median filter function based on DWT-SVD. In: International Conference Advanced Computing and Communication Technology, pp. 47–52 (2014, February)
9. Cox, I.J., Miller, M.L., Bloom, J.A., Fridrich, J., Kalker, T.: Digital Watermarking and Steganography. Morgan Kaufmann, San Mateo (2008)
10. Nezhadarya, E., Wang, Z.J., Ward, R.K.: Robust image watermarking based on multiscale gradient direction quantization. IEEE Trans. Image Process. **6**(4), 1200–1213 (2011)
11. Shukla, S.S.P.; Singh, S.P., Shah, K., Kumar, A.: Enhancing security & integrity of data using watermarking & digital signature. In: 2012 1st International Conference on Recent Advances in Information Technology (RAIT), pp. 15–17 (2012, March); Proceeding of IEEExplore, pp. 28–32. Print ISBN: 978-1-4577-0694-3
12. Shubhangi, D.C., Malipatil, M.: Authentication watermarking for transmission of hidden data using biometric technique. Int. J. Emerg. Technol. Adv. Eng. **2**(5) (2012). ISSN 2250-2459
13. Mardolkar, S.B., Shenvi, N.: A blind digital watermarking algorithm based on DWT-DCT transformation. Int. J. Innov. Res. Electr. Electron. Instrum. Control. Eng. **3**(Special Issue 2) (2016). ISSN (Online) 2321-2004
14. Sharma, D., Aggrawal, A., Gupta, A., Rai, H., Singh, N.: Int. J. Adv. Technol. **7**(3) (2016). ISSN:0976-4860
15. Parmar, P., Jindal, N.: Image security with integrated watermarking and encryption. IOSR J. Electron. Commun. Eng. **9**(3) (2014). p-ISSN: 2278-8735
16. Langelaar, G.C., Setyawan, I., Lagendijk, R.L.: Watermarking digital image and video data. IEEE Signal Process. Magaz. (2000)
17. Wolfgang, R.B., Podilchuk, C.I., Delp, E.J.: Perceptual watermarks for digital images and video. Proc. IEEE **87**(7), 1108–1126 (1999)
18. Potdar, V.M., Han, S., Chang, E.: A survey of digital image watermarking techniques. In: IEEE International Conference on Industrial Informatics (INDIA)-2005 (2005)

# Efficient Decentralized Key Management Approach for Vehicular Ad Hoc Network

Shikha Rathore, Jitendra Agrawal, Sanjeev Sharma and Santosh Sahu

## 1 Introduction

With the exponential growth of cars on highways throughout [1] latest years in India, and the great quantity of fatal injuries have allowed the researchers for the improvement of the latest technology to assist the drivers to travel more safely on highways. One primary purpose of traffic injuries is that drivers cannot always and correctly reply to the converting highway road conditions. In truth, maximum of the accidents may be prevented if drivers should gain relevant statistics of site visitors that is beyond their imagination and prescient, using vehicular communication technology fast growth and advancement of wireless technology create new avenues to utilize those technologies with the help of vehicular safety packages. The new Dedicated Short Range Communication (DSRC) or IEEE 802.11p protocol 11p protocol enables a newer class of vehicular safety applications which will increase the overall safety on the highway road, reliability, and efficiency of the current transportation system. Vehicular Ad hoc Networks (VANET), which is a part of Intelligent Transportation Systems (ITS), will provide a wide spectrum of applications to avoid highway accident.

Vehicular Ad hoc Networks (VANET), a part of Intelligent Transportation Systems (ITS) is referred to as the incorporated applications of the advanced technologies in Information Technology, good judgment controls and sensor networks provide

S. Rathore (✉) · J. Agrawal · S. Sharma · S. Sahu
School of IT, UTD, RGPV, Bhopal, India
e-mail: Shikharathore257@gmail.com

J. Agrawal
e-mail: jitendra@rgtu.net

S. Sharma
e-mail: sanjeev@rgtu.net

S. Sahu
e-mail: santoshsahu@rgtu.net

© Springer Nature Singapore Pte Ltd. 2019
R. K. Shukla et al. (eds.), *Data, Engineering and Applications*,
https://doi.org/10.1007/978-981-13-6351-1_13

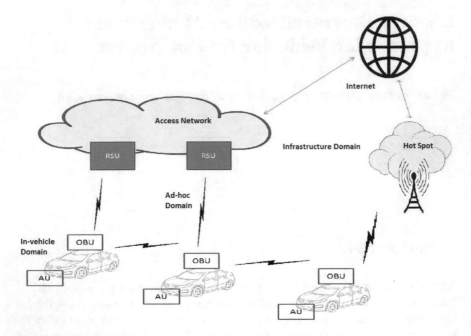

**Fig. 1** VANET architecture

travelers and authorities essential statistics they need to make the transportation
system extra secure, efficient, effective and reliable. On the grounds of the appearance
of vehicular ad hoc networks, lots of research work for real-time transportation
system management, has been carried out. Recent advances in wireless and sensor
technologies unexpectedly sell the seamless integration of numerous varieties of
data from transportation networks, to be gain drivers and offer a big selection of
transportation-orientated services. Vehicle-to-vehicle and vehicle-to-infrastructure
communications will be critical nearly in the close to future resulting in an operational
internet on the highways referred to as vehicular ad hoc networks (VANET) that will
revolutionize our idea of traveling [2] (Fig. 1).

## 2 Related Work

A review of the various techniques which used for key management in VANET is
presented in this section.

Hao et al. [3] have proposed OKD scheme which is a centralized group key
management protocol and uses periodic rekeying to decrease the communication
cost. The OKD method constructs the key tree with the unidirectional key approach
and derives the new key from the previous one to enhance the rekeying potency.

Guo et al. [4] Author proposed dispensed key management shape, which depends on short institution signature scheme to offer privates in VANET. Within the institution signature, public key of 1 organization is hooked up with non-public keys of more than 1 agency.

Guo et al. [5] have proposed the technique of binary search in RSU message verification section. Bloom filter out is used to update hash fee in notification message for lowering message overhead and improving the effectiveness of verification phase.

Daeinabi et al. [6] proposed concept of learning automata and Bayesian coalition game where each vehicle cooperate with one another to share information. Certificate authority supplies the key to vehicles and maintains certificate revocation list (CRL).

Kumar et al. [1] proposed an efficient decentralized public-key infrastructure (PKI) using the concepts of Bayesian Coalition Game (BCG) and Learning Automata (LA) are assumed as the players in the game, which coordinates with one another for information sharing.

# 3 Security Challenges in VANET

Safety demanding situations in VANET the demanding situations of security ought to be taken into consideration at some point of the layout of VANET structure, security protocols, and cryptographic algorithm, etc. The subsequent listing offers a few protection demanding situations [7]:

  I. **Real-Time Constraint**: Time is crucial in VANET where safety messages are to be added with a transmission postpone of less than 100 ms. A quick cryptographic algorithm needs to be used to reap real-time constraint. Message and entity need to be authenticated in real.

 II. **Data Consistency Liability**: Even an authenticated node can from time to time carry out malicious activities inflicting injuries or disturbance within the network. This inconsistency can be prevented by designing a mechanism which uses correlation amongst information received from diverse nodes on precise information.

III. **Low Tolerance for Error**: Protocols are typically designed for the use of opportunity algorithms. Statistics crucial to lifestyles are processed in very quick period of time in VANETs. In such cases, even a small mistake in the probabilistic set of rules can motive extraordinary harm.

IV. **Incentives**: Manufactures are interested in building programs preferred by way of consumers broadly speaking. Most effective are a small fraction of these customers may additionally agree having a vehicle that may record any traffic rule violation automatically. Consequently, incentives for automobile producers, purchasers and authorities pose an assignment to enforce protection in VANET.

 V. **High Mobility**: Because the nodes in VANET are incredibly cellular, for this reason a low execution time of protection protocols is wanted to attain same

throughput that stressed community produces. As a result, the security protocols have to be designed the usage of procedures to lessen the execution time.
Strategies can be implementing to fulfill this requirement.

VI. **Low-Complexity Security Algorithms**: Protection protocols like DTLS,
SSL/TLS, WTLS, use RSA primarily based public key cryptography. RSA
algorithm uses the integer factorization on large prime no., which is NP-
difficult. Accordingly, decryption of the messages encrypted the use of RSA
algorithm becomes very complicated and time consuming.
Consequently lattice based cryptosystems and remaining cryptographic set of
rules like Elliptic curve cryptosystems may be used to lessen this complexity.
AES can be used for bulk information encryption.

VII. **Transport Protocol Choice**: DTLS ought to be used preferably over TLS for
secure transaction over IP, because DTLS operates on connectionless transport
layer. Use of IPSec need to be avoided because it calls for too many messages
to set up. However, both IPSec and TLS may be used collectively when cars
are not in movement.

## 4 Security Requirements in VANET

VANET have to fulfill a few security necessities before they are deployed. A security
gadget in VANET should satisfy the following requirements [4]:

### 4.1 Authentication

Authentication ensures that the message is generated by the valid person. In VANET,
a vehicle reacts upon the statistics got here from the opposite vehicle as a result
authentication must be happy.

### 4.2 Availability

Availability requires that the statistics need to be available to the legitimate customers.
DoS Attacks can convey down the community and for this reason, facts cannot be
shared.

## 4.3 Non-repudiation

Non-repudiation approach a node cannot deny that he/she does not transmit the message. It may be essential to decide the suitable collection in crash reconstruction.

## 4.4 Privacy

The privacy of a node toward the unauthorized node should be guaranteed. That is required to get rid of the rubdown postpone assaults.

## 4.5 Data Verification

A regular verification of facts is required to remove the false messaging.

## 5 Types of Attackers in VANET

For securing the VANET, first we want to find out the forms of attackers, their nature of assaults, and their ability to harm the device. On the premise of capacity, those attackers can be of 3 types [4].

I. **Insider and Outsider**: Insiders are the existing true individuals of community having excessive ability whereas Outsiders are the intruders and that they have restrained capability to assault.

II. **Malicious and Rational**: Malicious attackers do not have any non-public advantage at the back of the assault; they simply want to damage the community's functionality. Rational attackers have a private advantage at the back of the attacks, therefore, they are predictable.

III. **Active and Passive**: Active attackers attack with the aid of generating indicators or packets while passive attackers only sense the community.

## 6 Types of Attacks in VANET

To achieve better protection from attackers, we must know about the types of attacks, which can be used against VANET's security [6] (Fig. 2):

Attacks on distinctive security requirement are given beneath [6]:

I. **Impersonate**: On this assault, the attacker assumes identity and privileges of a licensed node, either to apply network assets which are not to be had generally, or

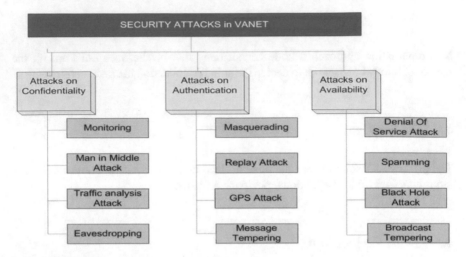

**Fig. 2** Classification of security attacks in VANET

with the rationale to disrupt the community's normal functioning. This assault is carried out by means of energetic attackers; they will be insider or outsiders. This is a multilayer assault, which means that the attacker can exploit the vulnerability of network layer, utility layer and/or transport layer. The attack may be done in two methods:

   (a) **False attribute possession**: On this, the attacker steals legitimate consumer's residences and uses its attributes to assert that it is for who (legitimate consumer) sent this message. Through the use of this type assault a regular car can claim that he/she is a police or fireplace protector to unfastened the visitors.

   (b) **Sybil**: In this attack, an attacker uses multiple identities at the same time.

 II. **Session Hijacking**: Most authentication methods are executed at the begin of the consultation. Therefore, it will become smooth for hackers to hijack the session after connection establishment. On this attack, the attackers take manipulate of session between nodes.

III. **Identity Revealing**: Normally a driver is itself owner of the vehicles subsequently getting proprietor's identity can placed the privacy at danger.

IV. **Location Tracking**: A vehicle's place at a given moment or the path followed by it for a time may be used to hint the car's present-day place and get records about the driving force.

 V. **Repudiation**: The principle chance in repudiation is denial or attempt and denial via a node involved in the verbal exchange. This is exceptional from the impersonate assault. in this attack two or more entity has not unusual identity subsequently it is simple to get indistinguishable and for this reason, they can be repudiated.

VI. **Eavesdropping**: It's far the most common assault type on confidentiality. This attack is made on the network layer and is passive in nature. Eavesdropping is achieved to get right of entry to personal user information.

VII. **Denial of Service**: It is the most common attack type on confidentiality. Right here the attacker prevents the valid user from the use of the services.

# 7 Group Key Management Approaches

Group key management approaches are classified into 3 major categories:

1. **Centralized [06]** on this methodology key distribution capability is carried with the aid of a single entity, which governs key generation and distribution whenever needed. Thus, a group key management protocol looks to attentive storage needs, computational force on both client and server sides,

2. **Distributed [04]** it is based totally on organization signature to make sure privacy in vehicular ad hoc network (VANETs). All the present organization signature schemes are based on a centralized key control wherein the keys are preloaded into the vehicle.

3. **Decentralized [05]**: This approach is a mixture of former approaches (centralized and distributed). An authorized cluster is split to some little subgroups so they create some graded levels. This approach shares the advantage and disadvantage that are expressed in centralized and distributed schemes. Considerately of those reasons, this theme is a lot of enticing for using during this work (Fig. 3).

**Fig. 3** Basic model for group key management

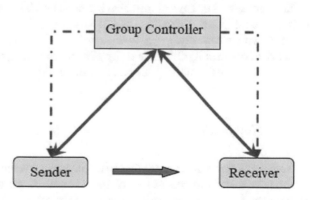

# 8  Mechanism of Group Key Management

## 8.1  Key Distribution

**Data Confidentiality**: Solely licensed members will access knowledge distributed among cluster members [8].
**Key Distribution**: All necessary keys ought to be distributed firmly before beginning communication.

The safety mechanisms implemented in VANET is depending on keys. Every message in VANET is encrypted, for this reason desires to be decrypted on the receiver's end either with the same key or exceptional key. Also one of a kind automobile manufacturers can deploy keys in exclusive methods. Additionally, belief on CA turns into fundamental difficulty, in public key infrastructure. Therefore, distributing keys among automobiles becomes a primary task in designing safety protocols [9].

## 8.2  Key Revocation

**Forward secrecy**: It requires that if a vehicle user UN agency have left the cluster can't access any future key of that cluster thus recent member can't decode knowledge when feat the cluster. To assure forward secrecy, a replacement cluster key 1 calculated when feat any cluster member.
**Key independence**: It ensures that the revelation of 1 key should not compromise alternative keys.
**Key revoked**: The keys of misbehaving nodes within the network are revoked and it is entered into key revocation list and as a result of newly calculated key should not match with recent key.
**Service Availability**: If single entity fails in key management then it should not have an effect on or forestall operation of whole multicast session.

## 8.3  Key Update

**Backward Secrecy**: It needs that if a replacement vehicle joins the cluster, it should not have access to recent key of that cluster. A member cannot decode knowledge communicated before connecting to that cluster. To assure backward secrecy, a replacement cluster key I is calculated when connecting a replacement cluster member.
**Local Rekeying**: If a member of any cluster joins or leaves the cluster then rekeying method of the subgroup should not have an effect on the whole cluster. This can be conjointly called one drawback.

**Fig. 4** Figure showing
central authority for two
groups in VANET

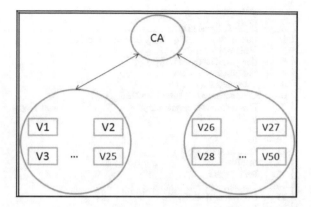

**Key Update**: If any key update is finished then cluster members ought to be notified regarding it.

**Computation requirements**: Because the wireless devices encounter with limitation in the process, they cannot stand intensive scientific discipline computation. The quantity of cryptography (decryption) ought to be unbroken as low as possible.

# 9 Proposed Schemes

The diagram below shows VANET clusters, each node having 25 vehicles. CA represents the certification authority that implements "Higher Than" rule for key distribution and management (Fig. 4).

**Key Distribution**:

**Multi-group Key Generation Using Master Key**

The algorithm is as follows: Consider r public–private key pairs. Let $e_i$ and $d_i$ be the ith public and private keys, respectively. Also, let $p_i$ and $q_i$ be their prime numbers, i.e., $e_i d_i = 1 \mod \varphi(p_i, q_i)$ where $\varphi(n)$ is Euler's quotient function. If $e_M$ and $d_M$ are the master keys, used for the encryption and decryption, the following congruence equations are established for any plaintext p and ciphertext r C:

$$P^{eM} = P^{ei}(mod\ p_iq_i)$$
$$C^{dM} = C^{di}(mod\ pi\ qi)$$

1.  Determine $p_1,\ldots,p_r$ , $q_1,\ldots,q_r$ from safe prime numbers.
2.  Start For i = 1 to r
3.  Compute $\Phi_i = (p_i - 1) \times (q_i - 1)$;
4.  Compute$x_i = (p_i - 1)/2$;
5.  Compute$y_i = (q_i - 1)/2$;
6.  Generate a random key$e_i = 4 \times Random + 1$;
7.  Now calculate decryption key$d_i = e_i^{2(x_i - 1)(y_i - 1)-1} \ mod\ 4(x_iy_i)$;
8.  End For
9.  n = 1;
10. For i = 1 to r
11. Compute$n = n \times (x_iy_i)$;
12. End For
13. For i = 1 to r
14. Compute$M[i] = n / (x_iy_i)$;
15. Compute$N[i] = M[i]^{(x_i - 1)(y_i - 1)-1} \ mod\ (x_iy_i)$;
16. End For
17. $e_m = 0$;
18. For i = 1 to r
19. Now calculate encryption key$e_m = (e_m + (e_i x\ M[i] \times N[i]\ ))\ mod\ n$;
20. End For
21. While($e_m$mod 4 != 1) $e_m = e_m + n$;
22. sleep;
23. Interrupt(when $j^{th}$ key pair should be updated)
24. Generate a random key$e_j = 4 \times Random + 1$;
25. Now calculate decryption key$d_i = e_i^{2(x_i - 1)(y_i - 1)-1} \ mod\ 4(x_iy_i)$;
26. go to 17;

---

**Algorithm_ Main**

{

1.  Configure mobile ad-hoc network for n nodes.
2.  From a pair of clusters of n/2 nodes **each**.
3.  Type a certification authority (CA) for key generation and Management.
4.  CA Generates multi cluster key exploitation master which can be used for digital signature.
5.  Distribute key pairs **to any or all** mobile nodes in networks
6. Encrypt using private key of the sender

}

## Key Revocation

**Algorithm node _join**

{

1.  Generate key pair for new node by using CA.
2.  Public key distribution for brand new node in cluster it joined.

}

**Key Update**

---

**Algorithm node_ leave**

{
  1. Key revocation of the node that had left network by CA.
  2. Inform all nodes in network so they'll take away that nodes public key from their list
     }

---

## 10 Flow Chart for Proposed Approach

A certification authority is formed that is positioned at the center of the road map. Keys generated by multi-cluster key management set of rules area unit distributed to all or any vehicles by CA. all the cars speak by suggests that of encrypting messages before causing and messages area unit decrypted by using the public key of the sender. In this chapter, the proposed scheme for key distribution and management had been mentioned for VANET [10] (Fig. 5).

## 11 Results

**Simulation Parameters**

| Parameter name | Value |
|---|---|
| Channel Type | Wireless |
| Propagation | Ground wave propagation |
| Mac Type | 802.11 |
| Antenna | Omnidirectional |
| Routing protocol | AODV |
| No of vehicle | 25 |

| Process | Time (ms) |
|---|---|
| Key generation time using ECC by RSUs | 814 |
| Key distribution to all 50 vehicles in simulation by respective RSUs | 396 |
| Time elapsed in encryption per packet | 176 |
| Time elapsed in decryption per packet | 171 |

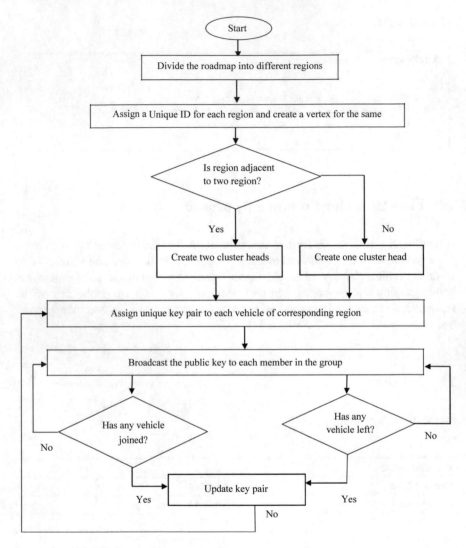

**Fig. 5** Flowchart for the proposed approach

**Fig. 6** Graph showing ID of
packets generated in
simulation

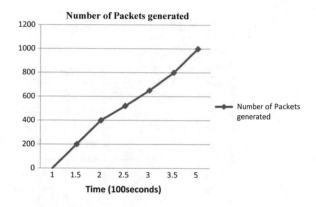

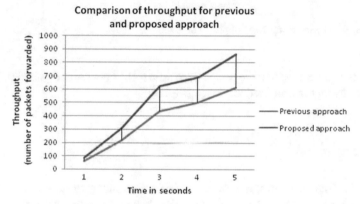

**Fig. 7** Graph showing comparison of throughput

| Number of Vehicles | Latency | | |
|---|---|---|---|
| | Bayesian approach | Decentralized | Proposed approach (s) |
| 10 | 0.07 | – | 0.035 |
| 20 | 0.097 | – | 0.04 |
| 30 | 0.11 | – | 0.071 |
| 40 | 0.23 | – | 0.084 |
| 50 | 0.31 | – | 0.09 |

In this graph, we see that at an average of $218-220$ id of the generated packet at a given node per/second using the proposed approach (Fig. 6).

On this graph, we see that the throughput or range of packets forwarded in step with 2nd has accelerated the use of the proposed approach. The quantity of packets forwarded consistent with second using proposed method is at a mean of $155-165$ packets in step with second (Fig. 7).

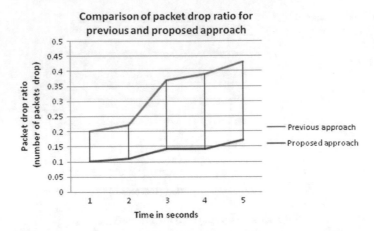

**Fig. 8** Graph showing comparison of packet drop ratio

In this graph, we see that the packet drop ratio (~0.11 per second) has significantly decreased using the proposed approach (Fig. 8).

## 12 Conclusion

VANET is a subclass of MANET, but it has a fast-changing ad hoc network and covers a large geographical area. So it requires key management scheme, which has a high throughput and is fast enough to handle quickly changing VANET ad hoc network. It is found that key distribution and proper key management protocol helps to increase the efficiency of the cryptography algorithms applied on to VANET. Key distribution and management is the backbone of VANET security. The proposed work uses multi-group key management protocol for key generation and encryption, and AES algorithm has been used for key decryption. Since the complexity of these algorithms is low, it ensures a high-speed key generation, encryption and decryption thus increasing the efficiency of VANET. Also, the packet drop ratio is greatly reduced due to fast and efficient key management scheme used in the proposed approach. The experimental results show that the use of decentralized key distribution mechanism is able to overcome the bottleneck situation and greatly increases the throughput for the key management system in the proposed approach.

## References

1. Kumar, N., Iqbal, R., Misra, S., Rodrigues, J.J.: An intelligent approach for building a secure decentralized public key infrastructure in VANET. J. Comput. Syst. Sci. **81**(6), 1042–1058

(2015). https://doi.org/10.1016/j.jcss.2014.12.016

2. Toor, Yasser, Muhlethaler, Paul, Laouiti, Anis, De La Fortelle, Arnaud: Vehicle ad hoc networks: applications and related technical issues. IEEE Commun. Surv. Tutor. **10**(3), 74–88 (2008)

3. Hao, Y., Cheng, Y., Zhou, C., Song, W.: A distributed key management framework with cooperative message authentication in VANETs. IEEE J. Sel. Areas Commun. **29**(3), 616–629 (2011). https://doi.org/10.1109/jsac.2011.110311

4. Guo, M.H., Liaw, H.T., Chiu, M.Y., Deng, D.J.: On decentralized group key management mechanism for vehicular ad hoc networks. Secur. Commun. Netw. (2012). https://doi.org/10.1002/sec.541

5. Guo, M.H., Liaw, H.T., Deng, D.J., Chao, H.C.: Centralized group key management mechanism for VANET. Secur. Commun. Netw. **6**(8), 1035–1043 (2013). https://doi.org/10.1002/sec.676

6. Daeinabi, A., Rahbar, A.G.: An advanced security scheme based on clustering and key distribution in vehicular ad-hoc networks. Comput Electr. Eng. **40**(2), 517–529 (2014). https://doi.org/10.1016/j.compeleceng.2013.10.003

7. Hartenstein, H., Laberteaux, L.P.: A tutorial survey on vehicular ad hoc networks. IEEE Commun. Mag. **46**(6), 164–171 (2008)

8. Verma, M., Huang, D.: SeGCom: secure group communication in VANET. In: 6th IEEE Conference In Consumer Communications and Networking, CCN, January 2009, pp. 1–5. https://doi.org/10.1109/ccnc.2009.4784943

9. Kushwaha, D., Shukla, P.K., Baraskar, R.: A Survey on Sybil attack in vehicular ad-hoc network. Int. J. Comput. Appl. **81**(6), 1042–1058 (2015). https://doi.org/10.5120/17262-7614

10. Lin, J.C., Lai, F., Lee, H.C.: Efficient group key management protocol with one-way key derivation. In: The IEEE Conference on Local Computer Networks, 30th Anniversary, November 2005, pp. 336–343. https://doi.org/10.1109/lcn.2005.61

# Image Forgery Detection: Survey and Future Directions

Kunj Bihari Meena and Vipin Tyagi

## 1  Introduction

One famous proverb says "A picture is worth a thousand words". Now everybody understands the essence of this idiom. But due to the availability of sophisticated tools for image manipulation, it is very easy to tamper the image by anyone with a modicum of computer skills. Hence, authenticity of image is challenged openly, therefore somewhere the above idiom loses its essence.

According to Merriam-Webster dictionary, digital image forgery is defined as "falsely and fraudulently altering a digital image". Image forgery is not a new concept; it started way back in 1840. French photographer Hippolyte Bayard created the first tampered image (Fig. 1) entitled with, "Self Portrait as a Drowned Man", in which, Bayard has professed to commit suicide [1].

More than a century ago, during American Civil War, a photo of American commanding general, General Ulysses S. Grant came into existence, which claimed that General Grant was sitting on horseback in front of his troops, at City Point, Virginia [2]. Later on, it has been found that image was not authentic; rather it was a composite of three images formed using negatives of the photographs.

Almost a decade ago, Iran has been accused of doctoring an image from its missile tests; the image [3] was released on the official website, Iran's Revolutionary Guard, which claimed that four missiles were heading skyward simultaneously. Recently, in July 2017, a fake image of Russian president Vladimir Putin was circulated over the social media related to the meeting with American president Donald Trump during the G20 summit 2017. This faked image garnered several thousand likes and retweets [4].

K. B. Meena · V. Tyagi (✉)
Jaypee University of Engineering and Technology, Raghogarh, Guna, MP, India
e-mail: dr.vipin.tyagi@gmail.com

© Springer Nature Singapore Pte Ltd. 2019
R. K. Shukla et al. (eds.), *Data, Engineering and Applications*,
https://doi.org/10.1007/978-981-13-6351-1_14

**Fig. 1** First fake image [1]

Image has remarkable role in various areas such as forensic investigation, criminal investigation, surveillance systems, intelligence system, sports, legal services, medical imaging, insurance claim, and journalism.

Substantial amount of research has been carried out in the last one decade in the field of forgery detection. Figure 2 shows the bar chart of a number of publications versus four types of image forgery detection techniques (copy-move, image splicing, resampling, retouching) for last two decades, over the years 1998–2017, collected from Google Scholar. Few observations from this bar chart are: startling growth has been seen in copy-move forgery detection in last one decade, and a significant focus is also given on image splicing detection in the last one decade over the first decade. However, less focus was given on retouching detection, one reason behind this may be that retouching is the least pernicious type forgery because generally, retouched images are not used for illegal purposes.

Forgery detection techniques are broadly categorized into two categories; active (non-blind, Fig. 3) and passive (blind) [5]. Active forgery detection techniques need some prior information about the image which may have been embedded in the image at the time of capturing the image or during image acquisition or later stages. Digital watermarking [6–8] and digital signature [9, 10] are the examples of active forgery detection techniques, and these approaches can be used to test the authenticity of the

**Fig. 2** Number of publications in the field of image forensics over the last two decades

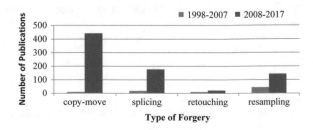

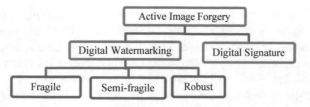

**Fig. 3** Active image forgery detection techniques

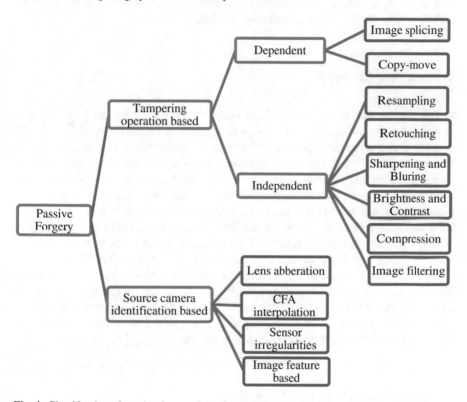

**Fig. 4** Classification of passive forgery detection techniques

image based on embedded information. On the basis of application, digital water-marking further can be categorized as fragile, semi-fragile, and robust watermarking [11]. In practicality, it is very rare that images produced for forensic investigation like fingerprint images, crime scene images, photographs of criminals, etc., would contain the watermark or signature, hence it can be concluded that active forgery detection techniques are not useful for forensic investigation of digital images.

On the contrary, passive forgery detection techniques do not need any prior knowl-edge about the image; rather these techniques identify manipulations by extracting intrinsic features of the image on the basis of the type of doctoring or photo-capturing device identification. Passive forgery further can be classified (Fig. 4) as dependent forgery and independent forgery. In dependent forgery, either tampering can be done in the same image by copying and pasting (cloning) [12] some area within the image or more than one image can be combined (image splicing) [13–15] to get convincing composite. On the other hand, independent forgery is the forgery in which some properties of the same image are manipulated. An example of independent forgery includes resampling, retouching, image rotation, scaling, resizing, addition of noise, blurring, image compression, etc. No involvement of prior knowledge about image makes passive forgery more practical in real life.

## 2  Existing Surveys

In the last decade, many surveys have been carried out on image forgery detection. Lanh et al. [16] discussed various techniques to detect image forgery based on camera. They have given conclusive remark that camera-based techniques are better than other forgery detection techniques, in terms of reliability. Farid [17], categorized image forgery tools into five groups, pixel-based techniques, format-based techniques, camera-based techniques, physically based techniques, and geometric-based. He has elaborated each method in detail. Recently, Warif et al. [18], reviewed copy-move forgery detection techniques. They have categorized copy-move forgery detection mainly into two classes: block-based and keypoint-based approach. Table 1 shows existing survey papers available on Google Scholar during 2007–2017.

Detailed classification of forgery detection methods is shown in Fig. 4, wherein blind forgery detection techniques have been categorized as tampering detection based and source camera identification based techniques. Tampering detection techniques have been discussed in this paper. For complete survey on source camera identification based techniques, readers may refer to [16, 23, 24].

## 3  Tampering Detection Techniques

In context to digital image, tampering means any manipulation or alteration in image to change its semantic meaning for illegal or unauthorized purposes. A tremendous amount of images are produced before digital image forensic for investigating whether the image is authentic (no alteration in semantic meaning of image) or tampered. Since photo-editing tools are becoming increasingly ubiquitous, anybody can tamper with the image and may use it with malicious intention. Figure 4 shows various tampering detection techniques.

In this section, four major types of tampering detection techniques (image splicing, copy-move forgery, image resampling, image retouching detection) are presented. Out of these four tampering detection techniques, first two are exploited for detecting

**Table 1** Existing survey papers available on Google scholar

| S. n. | Author(s) | Contribution | Observations |
|---|---|---|---|
| 1 | Lanh et al. [16] | Reviewed various techniques in digital camera image forensics | • Intrinsic features based methods of camera hardware are more reliable and better in terms of accuracy as compared to methods based on other camera software parts<br>• Camera identification methods outperform as compared to other forgery detection methods<br>• Hardware dependent characteristics such as aberration and CRF are potentially more reliable than methods based on scene content like lighting and image statistics |
| 2 | Farid [17] | Categorized the image forgery detection techniques into five groups (pixel-based, format-based, camera-based, physically based and geometric-based) | • Some forensic tools may not detect advanced forgeries but other forgery detection techniques are much reliable to challenge image fakery<br>• Due to the advancement of image manipulation tools, an arms race between the forger and forensic analyst is inescapable |
| 3 | Mahdian and Saic [19] | Reviewed various method based on blind image forgery | • Existing methods produce considerably higher false positive rates than which are reported in the existing papers<br>• Existing methods are not fully automated, need human interpretation<br>• Need to develop more reliable and robust methods |
| 4 | Christlein et al. [12] | Reviewed state of the art approaches pertaining to copy-move forgery | • Keypoint-based methods better than block-based methods in term of performance (execution time), however, block-based methods give better detection accuracy |
| 5 | Birajdar and Mankar [20] | Reviewed forgery detection techniques with more emphasis on passive tampering detection | • Existing methods are not automated, outputs need a human interpretation<br>• Existing methods are not effective when small regions are copy-moved<br>• Copy-move forgery detection, have shown high time complexity and false positives<br>• Need to extend forgery detection on audio and video |

(continued)

**Table 1** (continued)

| S. n. | Author(s) | Contribution | Observations |
|-------|-----------|--------------|--------------|
| 6 | Qazi et al. [21] | Surveyed various blind forgery detection techniques and classified into mainly three groups (copy-move, splicing, and retouching) | • DCT and PCA based techniques, exhibit high computational complexity and low accuracy rate<br>• DCT-based techniques are not effective when considering highly textured and small forged regions |
| 7 | Ansari et al. [22] | Various approaches of pixel-based image forgery detection have been reviewed | • Some algorithms are unable to detect forgery effectively. Some are having high time complexity<br>• Need to develop an efficient and accurate image forgery detection algorithm |
| 8 | Warif et al. [18] | CMFD techniques are organized into two approaches, namely: block-based and keypoint-based | • Existing techniques are not fit to solve real-world big data problems<br>• To increase processing speed, dimension reduction techniques like PCA, DWT, and SVD has been suggested<br>• Keypoint-based methods like SIFT and SURF are more reliable when geometrical transformation operations are taken into consideration |

dependent forgery and the last two tampering detection techniques are used for examining independent forgery.

## 3.1 Image Splicing Detection or Photo Composite Detection

In image splicing forgery, some part of image is copied and pasted on the other image to get forged image (Fig. 5). Image splicing is a basic step to create a photomontage from a set of images. To make composite image more realistic, postprocessing (scaling, cropping, retouching, rotating, etc.) operation may be applied on each of the components, furthermore, after performing splicing operation, again postprocessing operation can be applied to hide any discernible effects.

Although experts can identify image splicing forgery by just looking a forged image, in some cases. However, experienced forger can make composite image so elegant that it is almost impossible to say anything about the genuineness of an image, merely by looking at the image. When image splicing operation is carried out, some image statistics get disarranged. However, these statistical changes may not be perceptible to the human visual system. These statistic disarrangements of an image cannot be mitigated, even when expert burglar performs blending [25] and matting [26] operations on the forged image as a postprocessing operation.

Bicoherence features were proposed by Farid [27] to highlight the traces of tampered signal. In this paper, Farid has taken the assumption that in the frequency domain, a natural signal has weak higher order statistical correlations. Then after applying polyspectral analysis (bispectrum/bicoherence), he showed that "unnatural" correlations are introduced if the signal is passed through a nonlinearity. Farid has applied this technique to detect the splicing in human speech. Later on, in [28],

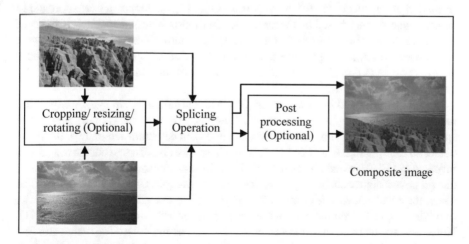

**Fig. 5** Process of image splicing

Ng et al. exploited bicoherence feature for detecting image splicing in images. They have proposed two methods, the first method exploits the dependence of the bicoherence features on the image content and the second offsets the splicing-invariant component from bicoherence. Detection accuracy of their method was from 62 to 70%. In their method, features were evaluated with Support Vector Machine (SVM). Later on, the same authors proposed [29] a new method to detect image spicing based on bipolar signal perturbation. They concluded that image splicing process increases the value of the bicoherence magnitude and phase features.

Considering lighting inconsistencies as a fundamental key feature for detecting image splicing, Farid et al. proposed a method for estimating the direction of an illuminating light [30]. Hilbert–Huang Transform (HHT) based technique was proposed in [31]. They have been exploiting SVM classifier and claimed 80.15% detection accuracy. Further work was carried out by Li et al. [32] and they used SVM classifier on moment features and HHT-based features together. They achieved detection accuracy of 85.87%, which is higher than that of the prior work (70% as reported in [28] and 80.15% in [31]) on the same evaluation dataset [33].

A natural image model was proposed by Shi et al. [34], in which statistical features extracted from the test image and 2D arrays were generated by applying multi-size block discrete cosine transform (MBDCT). The statistical features include moments of characteristic functions of wavelet subbands and Markov transition probabilities of difference 2D arrays. Dong et al. [35] devised a method for image splicing detection based on the statistical features extracted from image run length representation and image edge statistics. The support vector machine (SVM) is used as a classifier and achieved detection accuracy was 84.36%. Wang et al. [36], have given a new technique, using image chroma component. Hsu et al., in their research paper [13], presented a fully automatic method to detect splicing of digital images by incorporating three features: geometry invariant CRF estimation, consistency checking, and image segmentation. Unfortunately, the method was not robust enough and showed recall and precision about 70% only. Kakar et al. [37], developed a new approach to detect image forgery based on discrepancies in motion blur and spectral analysis of image gradients. They showed that their technique outperforms other contemporary techniques, which are applicable to motion blur. Carvalho et al. [14], designed a new method based on inconsistencies in the color of the illumination of image, by exploiting SVM meta-fusion classifier.

Rao et al. developed a new approach [38] to unveil splicing in image, by exploiting motion blur cues. Authors claimed that their approach can expose splicing even under space-variant blurring situations. A new method with detection accuracy of 98.2% to detect image splicing using Markov features in spatial and discrete cosine transform, invented by El-Alfy and Qureshi [39]. They further improved the performance of the proposed approach by applying the PCA (Principle Component Analysis) to select the most relevant features before building the detection model. Meanwhile, two other sophisticated methods were developed for unveiling splicing, in [40, 41]. Noise discrepancies in multiple scales are utilized as indicators for image splicing forgery detection in the paper [42] by Pun et al., they gave conclusive remark that their proposed method retains good detection accuracy in diverse situations like spliced area with different noise variance.

In [43], Park et al., introduced a method for image splicing detection, using the characteristic function moments for the inter-scale co-occurrence matrix in the wavelet domain with accuracy of 95.6%. They have tested their method on three popular datasets, Columbia, CASIA1, and CASIA2. Concurrently, Zhang and Lu [44], obtained an approach for unveiling image splicing by incorporating Markov model, in the block discrete cosine transform (DCT) domain and the Contourlet transform domain. In their illustrated method, authors have exploited SVM classifier to classify the authentic and spliced images for the gray image dataset. Verdoliva et al. devised an approach [45] by using autoencoder-based anomaly as a key feature.

Recently, Shen et al. [45] developed an algorithm for detecting image splicing by exploiting textural features based on the Gray-Level Co-occurrence Matrices. A support vector machine (SVM) is employed for classification purpose. The illustrated algorithm achieves the detection rates of 98% on CASIA v1.0, 97% on CASIA v2.0 and 91.88% on Columbia Image Splicing Detection Evaluation Dataset, with 96-D feature vector. Meanwhile, Farid [46] described three geometric techniques for detecting traces of digital manipulation in images. His proposed techniques were based on vanishing point, reflection, and shadow's location. Table 2 shows comparison of various algorithms for image.

## 3.2 Copy-Move Forgery Detection

In copy-move forgery one segment of image is copied and pasted in the other part of same image. Main intention of copy-move forgery is to hide some visual clues or replicating the things in image to mislead peoples. The prominent reason behind the surge in copy-move forgery is simplicity of this forgery. Good collection of tampered images throughout the history of image processing is available in [3]. Common workflow of copy-move forgery detection techniques has been shown in Fig. 6.

A survey on copy-move forgery detection techniques has been carried out in [18]. They have reviewed various research paper published in Web of Science (WOS) during years 2007–2014.

Silva et al. [53] developed a method for detecting copy-move forgery based on multiscale analysis and voting processes of a digital image. This method detects key points by exploiting Speeded-Up Robust Features (SURF) technique; Nearest Neighbor Distance Ratio (NNDR) is used for feature matching. Illustrated method can work under rotation, resizing or any combination of both. Unfortunately, it might not find a sufficient amount of key points in a small or homogeneous region.

Gabor filter based approach for copy-move forgery detection has been introduced by Lee et al. [54], which incorporates lexicographical sorting as a feature matching technique. Time complexity of this method was (O(PNlogN) + O(2JPN)). Meanwhile, in [55], Ardizzone et al. developed a copy-move forgery detection approach

**Table 2** Comparative study of existing techniques of image splicing detection

| S. n. | Algorithm | Features extracted | Classifier used | Accuracy | Dataset |
|---|---|---|---|---|---|
| 1 | Ng et al. [28] | Bicoherence features | SVM | 62–70% | CISDE |
| 2 | Fu et al. [31] | Hilbert–Huang Transform (HHT), and wavelet decomposition | SVM | 80.15% | CISDE |
| 3 | Shi et al. [34] | Moments of characteristic functions of wavelet subbands and Markov transition probabilities of difference 2-D arrays | SVM with RBF kernel | 91.87% | CISDE |
| 4 | Chen et al. [47] | 2D phase congruency and statistical moments of characteristic function | SVM | 82.32% | CISDE |
| 5 | Dong et al. [35] | Statistic moments of characteristic function of image run length histograms | SVM | 80.46–84.36% | CISDE |
|  | Wang et al. [36] | Image chroma component | SVM | 84.2% | CISDE |
| 6 | Li et al. [32] | HHT and moments feature | SVM | 85.87% | CISDE |
| 7 | Hsu and Chang [13] | Geometry invariant CRF estimation, consistency checking, and image segmentation | SVM | 70% precision, 70% recall | CUISDE |

(continued)

**Table 2** (continued)

| S. n. | Algorithm | Features extracted | Classifier used | Accuracy | Dataset |
|---|---|---|---|---|---|
| 8 | He et al. [48] | Approximate run length along with edge gradient direction | SVM with RBF kernel | 80.58% | CISDE |
| 9 | He et al. [49] | Markov features in DCT and DWT domain | SVM-RFE | 93.55% | CISDE and CASIA v1 |
| 10 | Carvalho et al. [14] | Inconsistencies in the color of the illumination of images | SVM meta-fusion | 85–86% | Dataset of 200 images taken from the Internet |
| 11 | Xu et al. [50] | The DCT Markov features in chroma channel | SVM | – | CUISDE |
| 12 | Qureshi et al. [39] | Markov features in spatial and discrete cosine transform, Principal Component Analysis (PCA) | SVM with RBF kernel | 98.2% | CISDE |
| 13 | Bahrami et al. [40] | Partial blur type inconsistency | Block-based partitioning | 94.6% | Dataset of 1200 tampered images |
| 14 | Zhao et al. [41] | 2D noncausal markov model | SVM | 93.36% | CISDE |
| 15 | Park et al. [43] | Characteristic function moments for the inter-scale co-occurrence matrix in the wavelet domain | SVM with RBF kernel | 95.3–95.6% | CASIA1 and CASIA2 |

(continued)

**Table 2** (continued)

| S. n. | Algorithm | Features extracted | Classifier used | Accuracy | Dataset |
|-------|-----------|--------------------|-----------------|----------|---------|
| 16 | Han et al. [51] | Markov feature | SVM | 94.87–98.50% | CASIA1 and CASIA2 |
| 17 | Zhang et al. [44] | Markov feature in block Discrete Cosine Transform (DCT) domain and the Contourlet transform domain | SVM-RFE | 96.69% | Dataset of 1150 forged color images |
| 18 | Rao and Ni [52] | Deep learning technique, and Convolutional Neural Network (CNN) | SVM | 96.38% | CASIA v1.0, CASIA v2.0 and CISDE |
| 19 | Shen et al. [45] | Textural features based on the gray-level co-occurrence matrices | SVM | 97–98% | CASIA v1.0 and CASIA v2.0 |

based on matching triangles, by applying mean vertex descriptors. This approach shows better performance in case of complex scenes; however, a lot of false matches occur with regular background.

Cozzolino et al. devised an algorithm [56] by considering dense-field techniques and Zernike moments as keypoint. Their algorithm utilizes nearest neighbor search algorithm and PatchMatch as a feature matching technique. Experiments were performed on copy-move forgery detection techniques in [57] by Li et al. and devised robust method for copy-move forgery detection by employing vlFeat software as feature extraction tool. Furthermore, Authors improved the accuracy of the obtained results by employing RANSAC via the gold standard algorithm. Their experiment showed the average precision as 0.86, however, method has high computation complexity, hence lead to low detection speed.

Pun et al. [58] proposed an algorithm to investigate copy-move forgery by combining keypoint feature and block-based feature. Experimental results show that their proposed scheme can achieve much better detection results for copy-move forgery images under various challenging conditions, such as geometric transforms, JPEG compression, and downsampling, compared with the existing contemporary copy-move forgery detection schemes. Also, they have measured precision value as 96.6%

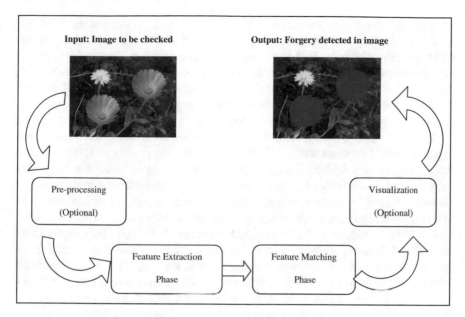

**Fig. 6** Common workflow of copy-move forgery detection techniques

and recall value as 100%. Block-based technique to detect copy-move forgery was given by Lee et al. [54], by using a histogram of orientated gradients, which can deal with images distorted by small rotations, blurring, adjustment of brightness, and color reduction. However, their approach fails if high rotation and scaling are introduced by forger.

A rotation-invariant method to detect copy-move forgery based on circular projection, was presented by Gürbüz et al. [59]. Meanwhile, Zhao et al. [60] proposed a technique to detect copy-move forgery by incorporating split-half recursion matching strategy to match SIFT keypoints. Method first calculates the affine transformation between two matched regions. And then, the ZNCC coefficients are measured to detect duplicate region.

Wenchang et al. introduced a new method [61] to detect copy-move forgery by employing new concept particle swam optimization (PSO) along with SIFT keypoint. In their experiment, authors have employed the best bin first (BBF) algorithm for feature matching. The method showed the precision of 99%, but unable to detect forgery when duplicated region is very small. Zandi et al. proposed a technique [62] based on interest point detector. In this paper, authors first detect all the interest points and then describe features using Polar Cosine Transform. After that, an improved version of the adaptive matching is employed. Furthermore, falsely matched pairs are discarded by an effective filtering algorithm. Moreover, to enhance the result, they have iterated process regarding the prior information. Authors claimed that their method can be exploited in other image processing areas, such as scene recognition or image retrieval, etc. However, method is vulnerable to resizing attack. Behavior knowledge space-based fusion was employed for copy-move forgery detection by Ferreira et al. [63]. In this work, they have overcome the limitations of fusing approaches by introducing new behavior knowledge space representation. Furthermore, authors also proposed multiscale behavior knowledge representation to deal with resizing and noise addition issues. Drawback with this method, however, is that it does not work well when image has several homogeneous regions.

Copy-move forgery detection technique based on scaled ORB has been proposed by Zhu et al. [64]. Their technique first, establishes a Gaussian scale space then, extracts FAST keypoints and the ORB features, in each scale space. Furthermore, technique employs RANSAC algorithm to remove the falsely matched keypoints. Experimental result shows that technique is effective for geometric transformation. However, approach is time-consuming when operated on high-resolution images. Bi et al. [65] designed a copy-move forgery detection technique by incorporating Multi-Level Dense Descriptor (MLDD) as a feature extraction method. They have utilized hierarchical feature matching method. Further, some morphological operations are applied to generate the final detected forgery regions. Approach work effectively with geometric transforms, JPEG compression, noise addition, and downsampling.

Ustubioglu et al. devised an algorithm for copy-move forgery detection by utilizing DCT-phase terms to restrict the range of the feature vector elements and also employed Benford's generalized law to determine the compression history of the image. The method uses element-by-element equality between the features. The method was also robust against postprocessing operations.

A new keypoint-based copy-move forgery detection for small smooth regions was developed by Wang et al. [66] by introducing the superpixel content based adaptive feature point detector. They also employed robust EMs-based keypoint features and fast Rg2NN based keypoint matching. However, serious limitation of method was the high computational complexity. Copy-move forgery detection technique has been devised by combining cellular automata(CA) and local binary patterns(LBP) by Tralic et al. [67]. The combination of CA and LBP allows a simple and reduced description of texture in the form of CA rules that represents local changes in pixel luminance values.

Recently in [68], a copy-move forgery detection method based on CMFD-SIFT, has been proposed by Yang et al. In their method, keypoints are detected by using a modified SIFT-based detector. This method improves the invariance to mirror transformation. Table 3 shows pros and cons of various algorithms for copy-move forgery detection based on several components such as feature extraction techniques, feature matching techniques, performance of algorithm and dataset used.

## 3.3  Image Resampling Detection

Resampling is mathematical technique to change the resolution (number of samples) of image, mainly for the purpose of increasing the size of image (upsampling) for printing banners and hoardings, etc., or for minimizing the size of image (downsam-

**Table 3** Comparative study of existing techniques of copy-move forgery detection techniques published between 2015 and 2017

| S. n | Algorithm | Feature extraction technique | Feature matching method | Performance | Pros/cons | Dataset |
|---|---|---|---|---|---|---|
| 1 | Silva et al. [53] | Multiscale analysis SURF | NNDR | CPU time 1.881 s/image | Pros: works under rotation, resizing, and these operations combined. Cons: not suitable for small or homogeneous region | CMH dataset CMEN datasets |
| 2 | Lee [69] | Gabor filter | Lexicographical sorting | O(PNlogN) + O(2JPN)) | Pros: work even when image is distorted by slight rotation and scaling, JPEG compression, blurring, and brightness adjustment | CoMoFoD dataset CMFDA |
| 3 | Ardizzone et al. [55] | Matching Triangles | Mean Vertex Descriptors | 10 s/image | Pros: better performance in case of complex scenes Cons: a lot of false matches image with regular background | CMFDA |

(continued)

**Table 3** (continued)

| S. n | Algorithm | Feature extraction technique | Feature matching method | Performance | Pros/cons | Dataset |
|------|-----------|------------------------------|-------------------------|-------------|-----------|---------|
| 4 | Cozzolino et al. [56] | Dense-field techniques and Zernike moments | nearest neighbor search algorithm and PatchMatch | 11 s/image | Pros: achieves higher robustness on rotations and scale changes Cons: slow performance | Database of 80 images along with [12] |
| 5 | Li et al. [57] | VlFeat software, RANSAC | K nearest neighbors | Precision is 86% | Pros: good detection accuracy Cons: slow performance | CMFDA along with MICC-F600 and MICC-F2000 |
| 6 | Pun et al. [58] | SIFT | Morphological operation | Precision 96.6% and recall 100% | Pros: better accuracy | CMFDA |
| 7 | Lee et al. [54] | Histogram of orientated gradients | Lexicographical sorting | Fc factor > 90% | Pros: robust against small rotations, blurring, adjustment of brightness, and color reduction Cons: not suitable with high rotation and scaling | CoMoFoD, second dataset of 30 high-resolution images |

(continued)

**Table 3** (continued)

| S. n | Algorithm | Feature extraction technique | Feature matching method | Performance | Pros/cons | Dataset |
|---|---|---|---|---|---|---|
| 8 | Gürbüz et al. [59] | Circular projection technique | Lexicographical sorted | Accuracy 99% | Pros: robust against scaling and mirroring operations. Cons: less effective while rotation angle is big | 20 test images with sizes of 326 × 245 |
| 9 | Zhao et al. [60] | SIFT | g2NN | – | Pros: effective with rotation, scaling and multiple copy operations Cons: accuracy decreases with compression | MICC F2000 |
| 10 | Wenchang et al. [61] | Particle Swam Optimization with SIFT | Best bin first | Precision 99% | Cons: fails to detect too small region | CMFDA |
| 11 | Zandi et al. [62] | Interest point detector | Adaptive matching | 436 ms/image | Pros: effective under various challenging conditions | SBU-CM161 |

(continued)

**Table 3** (continued)

| S. n | Algorithm | Feature extraction technique | Feature matching method | Performance | Pros/cons | Dataset |
|------|-----------|------------------------------|-------------------------|-------------|-----------|---------|
| 12 | Ferreira et al. [63] | BKS fusion | Multiscale Behavior Knowledge | $O(N^2)$ | Cons: less efficient, also fails when image has several homoge-neous regions, | CPH dataset |
| 13 | Zhu et al. [64] | FAST keypoints and the ORB features | hamming distance | 270 ms/image | Pros: effective for geometric transfor-mation Cons: time con-suming for forgery detection of high-resolution images. | Dataset of 107 real images and 107 tampered images |
| 14 | Bi et al. [65] | Multi-level Dense Descriptor | Hierarchical Feature Matching | F score > 91% | Pros: robust against various attacks | CMFDA dataset |
| 15 | Bi et al. [70] | Multiscale feature | Adaptive patch | F = 95.05% | Pros: good perfor-mance on downsam-pling and multiple copies | CMFDA dataset |
| 16 | Ustubioglu et al. [71] | DCT-phase term | Element-by-element equality | Accuracy 96% | Pros: robust against various postpro-cessing operations | Comofod and Kodak databases |

(continued)

**Table 3** (continued)

| S. n | Algorithm | Feature extraction technique | Feature matching method | Performance | Pros/cons | Dataset |
|------|-----------|------------------------------|-------------------------|-------------|-----------|---------|
| 17 | Wang et al. [66] | Superpixels classification and adaptive keypoints | Reversed g2NN | 221 s/image | Pros: effective with geometric transforms Cons: higher computational complexity | CMFDA dataset |
| 18 | Zheng et al. [72] | Zernike moments and SIFT along | g2NN | F = 84.91% | Pros: can detect smooth regions | CMFDA dataset CoMoFoD |
| 19 | Tralic et al. [67] | Cellular Automata (CA) and LBP | Euclidean distance | F > 0.92 | Pros: low computational complexity | CoMoFoD |
| 20 | Yang et al. [68] | Adaptive SIFT | AHC algorithm | F > 90% | Pros: improves the invariance to mirror transformation | CoMoFoD |
| 21 | Huang et al. [73] | FFT, SVD, and PCA | Exhaustive search | Accuracy 98% | Pros: high detection accuracy | CASIA v1.0 |

pling) for email and website use. In general, almost all sort of digital image forgery (more specifically image splicing) involve scaling, rotation or skewing operations to manipulate the image. In these operations use of resampling and interpolation processes is inevitable. Hence, it is possible to detect the image forgery by tracing the symptoms of resampling in image. Several papers have been published in the past decade to detect the forgery in image on the basis of resampling.

Popescu et al. proposed a method [74] to expose digital forgeries by detecting traces of resampling. In blind forgery detection, no prior information is available about image, like which particular postprocessing attack has been applied, which interpolation is used to resample the image or part of image. However, to identify traces of resampling, interpolation details might be a basic telltale cue of resampling detection. Hence, authors in exploited expectation/maximization algorithm

(EM) [75] to determine if a signal has been resampled. Two models were developed, one for those samples that are correlated to their neighbors, and the second model corresponds to those samples that are not correlated. Their method is effective to unveil the sign of linear or cubic interpolation. However, it fails to detect other more sophisticated nonlinear interpolation techniques.

Kirchner [76] introduced a method based on fixed linear predictor. Method extracts periodic artifact and detect resampling. Meanwhile, Mahdian and Saic [77] proposed an algorithm to detect interpolation and resampling with 100% detection accuracy. The method was based on derivative operator and radon transformation. Their method was effective to detect the traces of scaling, rotation, skewing transformations.

Li et al. developed an algorithm in [78] to detecting resampling based on periodicity introduced by resampling and JPEG compression. They employed EM algorithm to obtain the probability map of an image. Further, Fourier-transformed and matched with affine-transform templates employed to detect resampling. They have experimentally concluded that image is not undergone resampling if the periodicity of the probability map obtained. Moreover, they have examined their method on the dataset of 100 grayscale images and claimed the detection accuracy better than [74].

Lien et al. [79] illustrated a new approach to detect forgery by observing the detectable periodic distribution properties generated from the resampling and interpolation processes. Their approach divided resampling as horizontal and vertical and then applied detection technique. Experimentally, authors have claimed 95% detection accuracy of their method which in turn can verify one image of resolution 512 × 512 only in 50 s on their mentioned system. In [80] Qian et al. developed a method to detect blind image forgery using resampling history detection algorithm. Instead of calculating the exact resampling energy spectrum of second-order derivative rate, authors have proposed a special distance measurement for measuring how far apart two sub-images are away from each other in terms of resampling difference. Method can detect the resampling even when rotation has been performed after resampling.

In [81], Birajdar et al. invented a new technique that blindly detects global rescaling operation and estimates the rescaling factor based on the autocovariance sequence of zero-crossings of second difference of the tampered image. The method is robust to detect rescaling operation for images that have been subjected to various forms of attacks like JPEG compression and arbitrary cropping with accuracy of 99.5%.

Recently, David and Fernando [82] devised a new approach for the detection of resampling by incorporating new tools and concepts from RMT (Random Matrix Theory). RMT provides useful tools for modeling the behavior of the eigenvalues and singular values of random matrices. Striking positive aspect of the method was very low computational complexity. Meanwhile, Qian et al. also proposed a method for detecting resampling forgery in digital image by using linear parametric model. In their method, first resampling is detected in 1D signal then further they have extended it for 2D image.

Table 4 summarizes the pros and cons of various algorithms developed for resampling detection, based on several components such as feature extraction techniques, detection accuracy of algorithm, and dataset used.

**Table 4** Comparative study of existing techniques of resampling detection

| S. n. | Algorithm | Feature description | Detection accuracy/performance | Pros/cons | Dataset |
|---|---|---|---|---|---|
| 1 | Popescu and Farid [74] | EM algorithm | Accuracy 80% | Pros: work with GIF format also Cons: fail to detect other more sophisticated nonlinear interpolation | Database of 200 grayscale images in TIFF format with 512 × 512 size cropped from a smaller set of 25, 1200 × 1600 images |
| 2 | Kirchner [76] | Fixed linear predictor | Accuracy 100% for upsampling | Pros: fast and reliable | Database of 200 uncompressed 8 bit grayscale with resolution 3112 × 2382 pixels |
| 3 | Mahdian and Saic [77] | Derivative operator and radon transformation | 100% | Pros: capable of detecting traces of scaling, rotation, skewing transformations | Dataset of 40 images corrupted by various transformations |
| 4 | Wang and Ping [83] | Singular value decomposition | 79.838% | Pros: robust against scaling manipulation Cons: less accurate with rotating transformation and compression | UCID |
| 5 | Li et al. [78] | EM algorithm, Fourier transform and affine transform | Better detection accuracy than [74] | Pros: better on resampling detection in JPEG compression Cons: very time consuming | Database of 100 gray-level images of various resolutions and formats (TIFF, BMP, PNG etc.) |

(continued)

**Table 4** (continued)

| S. n. | Algorithm | Feature description | Detection accuracy/performance | Pros/cons | Dataset |
|---|---|---|---|---|---|
| 6 | Lien et al. [79] | Pre-calculated resampling weighting table | Accuracy 95% and CPU time 50 s/image | Pros: better detection accuracy | Dataset with 160 gray images with resolution 512 × 512 |
| 7 | Qian et al. [80] | DFT | 0.5203 s/image | Pros: effective with rotation after resampling | Dataset of 500 images cropped with different resampling rates |
| 8 | Feng et al. [84] | SVM | 100% | Pros: shows better performance for downsampling | BOSS database |
| 9 | Hou et al. [85] | Local linear transform | 96.15–98.75% | Pros: good resampling detection performance | Dataset of 1000 colored bmp images cropped into 512 × 512 pixels |
| 10 | Birajdar and Mankar [81] | Autocovariance sequence, DFT | 99.5% | Pros: work well with various forms of attacks like JPEG compression and arbitrary cropping | UCID dataset USC-SIPI dataset |
| 11 | David and Fernando [82] | Asymptotic eigenvalue distribution and Random Matrix Theory (RMT) | 0.0066 s/image | Pros: Low computational complexity | Dresden Image Database of a total of 1317 raw images [86] |
| 12 | Qiao et al. [87] | Probability of residual noise and LRT detector | 0.0996 s/image | Pros: effective with uncompressed/compressed non-resampled images | 500 uncompressed non-resampled images and 500 compressed resampled JPEG images with Quality Factor (QF) from 50 to 90 |

(continued)

**Table 4** (continued)

| S. n. | Algorithm | Feature description | Detection accuracy/performance | Pros/cons | Dataset |
|---|---|---|---|---|---|
| 13 | Su et al. [88] | Inverse filtering process with blind deconvolution | 90% | Pros: does not affect with JPEG block artifacts Cons: Not effective to detect blurred images | UCID |
| 14 | Peng et al. [89] | AR coefficients and normalized histograms | 98.3% | Cons: performance degrades with increasing JPEG compression ratio | BOSS dataset |
| 15 | Bayar and Stamm [90] | Convolutional Neural Network (CNN) | 91.22% | Pros: can detect resampling in recompressed images | Dataset of 6500 images of size at least 2688 × 1520 |

## 3.4 Image Retouching Detection

Retouching can be defined as "polishing of an image". In general, retouching refers to subsequently improving the surface of an image. Contrast enhancement is a widely used technique to remove obvious visual clues from the forged image as a postprocessing operation. However, more involvement of retouching can be seen in entertainment media, magazine covers, etc., where retouching is not used maliciously. Contrast enhancement operations are tantamount to pixel value mappings, which introduce some statistical traces [91]. Therefore, retouching can be exploited as a tool for image forgery detection.

Stamm et al. proposed a method [92] for detecting contrast enhancement in an image on the basis of gray value histogram. They have developed a model for the histogram of an unaltered image and then exploited this model to detect manipulated artifacts. Detection accuracy of the algorithm was claimed to be about 99%. Cao et al. [93] developed a technique to detect sharpening alteration in digital images. Authors have measured gradient aberration of the gray histogram generated from unsaturated luminance regions of an image and exploited to unveil traces of sharpening manipulation. Cao et al. [94] proposed a new method to detect unsharp masking sharpening based on the feature overshoot artifacts occurred around side-planar edges. By experimental study, authors claimed, their method to be accurate to detect sharpening on small size images even when post-JPEG compression and noising attacks employed. Same authors further explained a method [95] for detecting the contrast enhancement in digital images. This time, they have utilized the histogram peak/gap artifacts feature to detect global contrast enhancement applied to the previously JPEG-compressed images. Their proposed method is effective for detecting forgery when contrast enhancement is employed as the last step of manipulation. However, method fails to detect forgery when image is highly compressed.

In [91], Lin et al. explained that the contrast enhancement can disturb the interchannel similarities of high-frequency components, and then proposed a new method to detect the cut past forgery by detecting symptoms of contrast enhancement. Unfortunately, this method also fails when image is compressed after forgery. Ding et al. [96], proposed a new method to detect the special characteristic of the texture modification caused by the USM sharpening by employing edge perpendicular binary coding.

Recently, Zhu et al. presented a new approach [97] to detect image sharpening operation based on the overshoot artifact metric. First, they have detected edges using canny operator, then, non-subsampled contourlet transform (NSCT) is employed to classify image edge points. In the final stage, they have measured the overshoot artifact for each edge points and then, on the basis of overshoot artifacts judgment were made on sharpening operation. Table 5 shows various algorithms for retouching detection, based on several components such as; feature extracted, classifier applied, detection accuracy, and dataset used for testing the algorithm.

## 4  Datasets Available

Table 6 shows several publicly available datasets, which are frequently used by researchers.

## 5  Conclusion and Future Directions

In this paper, various existing methods on blind image forgery detection are reviewed. A broad classification of image forgery detection techniques is given. More specifically, a comprehensive overview of four main types of forgery detection techniques such as image splicing, copy-move, resampling, and retouching detection is given. Various existing methods have been reviewed in each category and observed that existing techniques suffer from one or more following limitations. (1) Detection accuracy (2) High computation complexity (3) Vulnerable against various attacks such as rotation, scaling, JPEG compression, blurring, and brightness adjustment, etc. (4) A lot of false matches with regular background.

Apart from abovementioned limitations, one major issue of these detection techniques is the limited scope of utilization, for example, method developed for copy-move forgery cannot work with image splicing or resampling and vice versa. In

**Table 5** Comparative study of existing techniques of retouching detection

| S. n. | Algorithm | Extracted feature | Classifier | Detection accuracy | Dataset used |
|---|---|---|---|---|---|
| 1 | Stamm and Ray [92] | Gray value histogram | Thresholding classifier | Global contrast 99%, Local contrast 98.5%, Histogram equalization 99% | 341 images captured using different digital cameras |
| 2 | Cao et al. [93] | Ringing artifacts | Fisher linear classifier | Precision 0.85 | Dataset of 403 JPEG images |
| 3 | Cao et al. [94] | Overshoot strength | Thresholding classifier | 88% | Dataset of 400 JPEG images with the size from 1200 × 900 to 2832 × 2128 pixels |
| 4 | Lin et al. [91] | Interchannel correlation | Thresholding classifier | 90% | Dataset of 100 uncompressed color images of size 1600 × 1200 |
| 5 | Cao et al. [95] | Histogram peak/gap artifacts | Thresholding classifier | 100% | BOSS public dataset and UCID |
| 6 | Ding et al. [96] | Rotation-invariant LBP | SVM | 90% | UCID |
| 7 | Zhu et al. [97] | Multiresolution overshoot artifact | NSCT | 92% | UCID |

spite of burgeoning research in the field of image forgery detection, no detection method can be used as a solution for detecting all kind of forgeries. Hence, there is a great need to develop a robust, sophisticated forgery detection technique which could eliminate aforementioned limitations. Furthermore, researchers may extend these techniques to detect forgeries in videos.

**Table 6** Description of various available datasets related to forgery detection

| S. n | Dataset | Forgery type | Total images | Resolution | Description |
|------|---------|--------------|--------------|------------|-------------|
| 1 | CISDE [33] | Splicing | 1845 | 128 × 128 | Contains 933 forged images and 912 authentic images, all are gray images in PNG format |
| 2 | CUISDE [98] | Splicing | 361 | 757 × 568, 1152 × 768 | Contains 180 forged images and 181 authentic images, all are colored images in TIFF format |
| 3 | CASIA v1.0 [99] | Splicing | 1725 | 324 × 256 | Contains 925 forged images and 800 authentic images, all are colored images in JPEG format |
| 4 | CASIA v2.0 [100] | Splicing | 12614 | 240 × 160–900 × 600 | Contains 5123 forged images and 7491 authentic images, all are colored images in JPEG format Also contains uncompressed images and JPEG images with different Q factors |
| 5 | CMFDA [12] | Copy-move | 48 | 420 × 300–3888 × 2592 | Contains original and forged image applied with JPEG compression, rotation and scaling operation |

(continued)

**Table 6** (continued)

| S. n | Dataset | Forgery type | Total images | Resolution | Description |
|------|---------|--------------|--------------|------------|-------------|
| 6 | CoMoFoD dataset [101] | Copy-move | 260 | 512 × 512–3000 × 2000 | Contains original and forged images, applied with translation, rotation, scale, distortion or a combination of them |
| 7 | MICC-F600 [102] | Copy-move | 600 | 800 × 533–3888 × 2592 | Contains original and forged images, that are randomly taken from MICC-F2000 and SATS-130 datasets |
| 8 | MICC-F2000 [103] | Copy-move | 2000 | 2018 × 1536 | Contains original and forged image, applied with translation, rotation, scale |
| 9 | SBU-CM161 [104] | Copy-move | 240 | 800 × 580 | Contains images based on 16 original JPEG images with rotation, scaling, compression |
| 10 | CPH [53] | Copy-move | 216 | 845 × 634–296 × 972 | Contains images with forgeries created through mixed operations such as resizing, rotation, scaling, compression, illumination matching |
| 11 | SCUT-FBP [105] | Retouching | 500 | 384 × 512 | Contains 500 different female face images along with the attractiveness rating scores computed from individual scores from 70 observers |

(continued)

**Table 6** (continued)

| S. n | Dataset | Forgery type | Total images | Resolution | Description |
|------|---------|--------------|--------------|------------|-------------|
| 12 | BOSS public dataset [106] | Retouching | 800 | 2000 × 3008–5212 × 3468 | Contains unaltered photograph images in raw format |
| 13 | UCID [107] | Retouching | 1338 | 384 × 512 | Contains uncompressed images in TIFF format on various topics such as natural scenes, man-made objects, indoors and outdoors |

# References

1. Revolvy.com: Hippolyte Bayard (French, 1801–1887). https://www.revolvy.com/topic/ Hippolyte Bayard&item_type = topic
2. Loc.gov: Civil War Glass Negatives and Related Prints. https://www.loc.gov/pictures/ collection/cwp/mystery.html (2008)
3. Photo Tampering Throught History. http://pth.izitru.com/
4. Tait, A.: How a badly faked photo of Vladimir Putin took over Twitter. http://www. newstatesman.com/science-tech/social-media/2017/07/how-badly-faked-photo-vladimir-putin-took-over-twitter (2017)
5. Tyagi, V.: Understanding Digital Image Processing. CRC Press (2018). ISBN 9781315123905
6. Wang, S., Zheng, D., Zhao, J., Tam, W.J., Speranza, F.: An image quality evaluation method based on digital watermarking. IEEE Trans. Circuits Syst. Video Technol. **17**, 98–105 (2007)
7. Singh, P., Chadha, R.S.: A survey of digital watermarking techniques, applications and attacks. IEEE Int. Conf. Ind. Inform. **2**, 165–175 (2013)
8. Arnold, M., Schmucker, M., Wolthusen, S.D.: Techniques and Applications of Digital Watermarking and Content Protection. A Cataloging in Publication Record, Artech House Inc, Norwood, MA, USA (2003)
9. Lu, C., Liao, H.M., Member, S.: Structural digital signature for image authentication: an incidental distortion resistant scheme. IEEE Trans. Multimed. **5**, 161–173 (2003)
10. Schneider, M., Chang, S.: A robust content based digital signature for image authentication. In: IEEE International Conference on Image Processing. pp. 227–230 (1996)
11. Cox, I.J., Miller, M.L., Bloom, J.A., Kalker, T.: Digital Watermarking and Steganography Second Edition
12. Christlein, V., Riess, C.C., Jordan, J., Riess, C.C., Angelopoulou, E.: An evaluation of popular copy-move forgery detection approaches. IEEE Trans. Inf. Forensics Secur. **7**, 1841–1854 (2012)
13. Hsu, Y., Chang, S.: Camera response functions for image forensics: an automatic algorithm for splicing detection. IEEE Trans. Inf. Forensics Secur. **5**, 816–825 (2010)
14. Carvalho, T.J.De, Member, S., Riess, C., Member, A., Angelopoulou, E., Pedrini, H., Rocha, A.D.R.: Exposing digital image forgeries by illumination color classification. IEEE Trans. Inf. Forensics Secur. **8**, 1182–1194 (2013)

15. Popescu, A.C., Farid, H.: Exposing digital forgeries by detecting traces of resampling. IEEE Trans. Inf. Forensics Secur. **53**, 758–767 (2005)
16. Lanh, T.V.L.T., Van Chong, K.-S., Chong, K.-S., Emmanuel, S., Kankanhalli, M.S.: A survey on digital camera image forensic methods. In: 2007 IEEE International Conference on Multimedia and Expo, pp. 16–19 (2007)
17. Farid, H.: A survey of image forgery detection techniques. IEEE Signal Process. Mag. **26**, 16–25 (2009)
18. Warif, N.B.A., Wahab, A.W.A., Idris, M.Y.I.: Copy-move forgery detection: survey, challenges and future directions. J. Netw. Comput. Appl. (2016)
19. Mahdian, B., Saic, S.: A bibliography on blind methods for identifying image forgery. Signal Process. Image Commun. **25**, 389–399 (2010)
20. Birajdar, G.K., Mankar, V.H.: Digital image forgery detection using passive techniques: a survey. Digit. Investig. **10**, 226–245 (2013)
21. Qazi, T., Hayat, K., Khan, S.U., Madani, S.A., Khan, I.A., Kołodziej, J., Li, H., Lin, W., Yow, K.C., Xu, C.-Z.: Survey on blind image forgery detection. Image Process. IET. **7**, 660–670 (2013)
22. Ansari, M.D., Ghrera, S.P., Tyagi, V.: Pixel-based image forgery detection: a review. IETE J. Educ. **55**, 40–46 (2014)
23. Ali, M., Deriche, M.: A bibliography of pixel-based blind image forgery detection techniques. Signal Process. Image Commun. **39**, 46–74 (2015)
24. Lukas, J., Fridrich, J., Goljan, M.: Detecting digital image forgeries using sensor pattern noise. In: Proceedings of SPIE, vol. 6072, pp. 60720Y–60720Y–11 (2006)
25. Yatziv, L., Sapiro, G.: Fast image and video colorization using chrominance blending. IEEE Trans. Image Process. 1120–1129 (2006)
26. Chuan, Y.Y., Curless, B., Salesin, D.H., Szeliski, R.: A bayesian approach to digital matting. Comput. Vis. Pattern Recognit. (2001)
27. Farid, H.: Detecting digital forgeries using bispectral analysis. Mit Ai Memo Aim-1657 Mit (1999)
28. Ng, T., Chang, S., Sun, Q.: Blind detection of photomontage using higher order statistics. In: IEEE International Symposium on Circuits System, pp. 7–10 (2004)
29. Ng, T., Chan, S.F.: A model of Image Splicing. In: IEEE International Conference on Image Process (2004)
30. Johnson, M.K., Farid, H.: Exposing digital forgeries by detecting inconsistencies in lighting. In: Proceedings of 7th Workshop on Multimed Security—MM&Sec'05, pp. 1–10 (2005)
31. Fu, D., Shi, Y.Q., Su, W.: Detection of image splicing based on Hilbert-Huang transform and moments of characteristic functions. Int. Work. Digit. Watermarking 177–187 (2006)
32. Li, X., Jing, T., Li, X.: Image splicing detection based on moment features and Hilbert-Huang transform. IEEE Int. Conf. Inf. Theory Inf. Secur. (2010)
33. Columbia DVMM Research Lab,Image Splicing Detection Evaluation Dataset. www.ee.columbia.edu/dvmm/researchProjects/AuthenticationWatermarking/Blind (2004)
34. Shi, Y.Q., Chen, C., Chen, W.: A natural image model approach to splicing detection. In: Proceedings of 9th Workshop Multimedia Security, pp. 51–62 (2007)
35. Dong, J., Wang, W., Tan, T., Shi, Y.Q.: Run-length and edge statistics based approach for image splicing detection. IWDW Int. Work. Digit. Watermarking. 5450 LNCS, 76–87 (2009)
36. Wang, W., Dong, J., Tan, T.: Effective image splicing detection based on image chroma. In: IEEE International Conference on Image Process, pp. 1257–1260 (2009)
37. Kakar, P., Member, S., Sudha, N., Member, S., Ser, W., Member, S.: Exposing digital image forgeries by detecting discrepancies in motion blur. IEEE Trans. Multimed. **13**, 443–452 (2011)
38. Rao, M.P., Rajagopalan, A.N., Member, S.: Harnessing motion blur to unveil splicing. IEEE Trans. Inf. Forensics Secur. **9**, 583–595 (2014)
39. El-Alfy, E.S., Qureshi, M.A.: Combining spatial and DCT based Markov features for enhanced blind detection of image splicing. Pattern Anal. Appl. 18, 713–723 (2015)

40. Bahrami, K., Member, S., Kot, A.C., Li, L., Li, H., Member, S.: Blurred image splicing localization by exposing blur type inconsistency. IEEE Trans. Inf. Forensics Secur. **6013**, 1–10 (2015)
41. Zhao, X., Wang, S., Li, S., Li, J.: Passive image-splicing detection by a 2-D noncausal Markov model. IEEE Trans. Circuits Syst. Video Technol. **25**, 185–199 (2015)
42. Pun, C.M., Liu, B., Yuan, X.C.: Multi-scale noise estimation for image splicing forgery detection. J. Vis. Commun. Image Represent. **38**, 195–206 (2016)
43. Park, T.H., Han, J.G., Moon, Y.H., Eom, I.K.: Image splicing detection based on inter-scale 2D joint characteristic function moments in wavelet domain. EURASIP J. Image Video Process **30** (2016)
44. Zhang, Q., Lu, W.: Joint image splicing detection in DCT and contourlet transform domain. J. Vis. Commun. Image Represent. (2016)
45. Shen, X., Shi, Z., Chen, H.: Splicing image forgery detection using textural features based on the grey level co-occurrence matrices. IET Image Process. **11**, 44–53 (2017)
46. Farid, H.: How to Detect Faked Photos (2017)
47. Chen, W., Shi, Y.Q., Su, W.: Image splicing detection using 2-D phase congruency and statistical moments of characteristic function. In: Security Steganography and Watermarking Multimedia Contents IX, vol. 6505, pp. 1–8 (2007)
48. He, Z., Sun, W., Lu, W., Lu, H.: Digital image splicing detection based on approximate run length. Pattern Recognit. Lett. **32**, 1591–1597 (2011)
49. He, Z., Lu, W., Sun, W., Huang, J.: Digital image splicing detection based on Markov features in DCT and DWT domain. Pattern Recognit. **45**, 4292–4299 (2012)
50. Xu, B., Liu, G., Dai, Y.: Detecting image splicing using merged features in chroma space. Sci. World J. (2014)
51. Han, J.G., Park, T.H., Moon, W.H., Eom, I.K.: Efficient Markov feature extraction method for image splicing detection using maximization and threshold expansion. J. Electron. Imaging. (2016)
52. Rao, Y., Ni, J.: A deep learning approach to detection of splicing and copy-move forgeries in images. In: 8th IEEE International Workshop Information Forensics Security WIFS (2016)
53. Silva, E., Carvalho, T., Ferreira, A., Rocha, A.: Going deeper into copy-move forgery detection: Exploring image telltales via multi-scale analysis and voting processes. J. Vis. Commun. Image Represent. **29**, 16–32 (2015)
54. Lee, J.C., Chang, C.P., Chen, W.K.: Detection of copy-move image forgery using histogram of orientated gradients. Inf. Sci. (Ny) **321**, 250–262 (2015)
55. Ardizzone, E., Bruno, A., Mazzola, G.: Copy-move forgery detection by matching triangles of keypoints. IEEE Trans. Inf. Forensics Secur. **10**, 2084–2094 (2015)
56. Cozzolino, D., Poggi, G., Verdoliva, L.: Efficient dense-field copy-move forgery detection. IEEE Trans. Inf. Forensics Secur. **10**, 2284–2297 (2015)
57. Li, J., Li, X., Yang, B., Sun, X.: Segmentation-based image copy-move forgery detection scheme. IEEE Trans. Inf. (2015)
58. Pun, C., Member, S., Yuan, X., Bi, X.: Oversegmentation and feature point matching. IEEE Trans. Inf. Forensics Secur. **10**, 1705–1716 (2015)
59. Gürbüz, E., Ulutaş, G., Ulutaş, M.: Rotation invariant copy move forgery detection method. In: Proceedings of 9th International Conference on Electrical and Electronics Engineering, pp. 202–206 (2015)
60. Zhao, F., Zhang, R., Guo, H., Zhang, Y.: Effective digital image copy-move location algorithm robust to geometric transformations. In: IEEE International Conference on Signal Processing, Communications and Computing (ICSPCC) (2015)
61. Wenchang, S.H.I., Fei, Z., Bo, Q.I.N., Bin, L.: Improving image copy-move forgery detection with particle swarm optimization techniques. China Commun. 139–149 (2016)
62. Zandi, M., Mahmoudi-Aznaveh, A., Talebpour, A.: Iterative copy-move forgery detection based on a new interest point detector. IEEE Trans. Inf. Forensics Secur. **11**, 2499–2512 (2016)

63. Ferreira, A., Felipussi, S.C., Alfaro, C., Fonseca, P., Vargas-Munoz, J.E., Dos Santos, J.A., Rocha, A.: Behavior knowledge space-based fusion for copy-move forgery detection. IEEE Trans. Image Process. **25**, 4729–4742 (2016)
64. Zhu, Y., Shen, X., Chen, H.: Copy-move forgery detection based on scaled ORB. Multimed. Tools Appl. **75**, 3221–3233 (2016)
65. Bi, X., Pun, C.M., Yuan, X.C.: Multi-level dense descriptor and hierarchical feature matching for copy-move forgery detection. Inf. Sci. (Ny) **345**, 226–242 (2016)
66. Wang, X., Li, S., Liu, Y.: A new keypoint-based copy-move forgery detection for small smooth regions. Multimed. Tools Appl. (2016)
67. Tralic, D., Grgic, S., Sun, X., Rosin, P.L.: Combining cellular automata and local binary patterns for copy-move forgery detection. Multimed. Tools Appl. 16881–16903 (2016)
68. Yang, B., Sun, X., Guo, H., Xia, Z., Chen, X.: A copy-move forgery detection method based on CMFD-SIFT. Multimed. Tools Appl. (2017)
69. Lee, J.C.: Copy-move image forgery detection based on Gabor magnitude. J. Vis. Commun. Image Represent. **31**, 320–334 (2015)
70. Bi, X.L., Pun, C.M., Yuan, X.C.: Multi-scale feature extraction and adaptive matching for copy-move forgery detection. Multimed. Tools Appl. 1–23 (2016)
71. Ustubioglu, B., Ulutas, G., Ulutas, M., Nabiyev, V.V.: A new copy move forgery detection technique with automatic threshold determination. AEU Int. J. Electron. Commun. **70**, 1076–1087 (2016)
72. Zheng, J., Liu, Y., Ren, J., Zhu, T., Yan, Y., Yang, H.: Fusion of block and keypoints based approaches for effective copy-move image forgery detection. Multidimens. Syst. Signal Process. **27**, 989–1005 (2016)
73. Huang, D., Huang, C., Hu, W.: Robustness of copy-move forgery detection under high JPEG compression artifacts. Multimed. Tools Appl. **76**(1), 1509–1530 (2017)
74. Popescu, A.C., Farid, H.: Exposing Digital Forgeries by Detecting Traces of Resampling Resampling Detecting Resampling Experiment Results (2005)
75. Dempster, A., Laird, N., Rubin, D.: Maximum lilelihood from in- complete data via the EM algorithm. J. Roy. Stat. Soc. **99**, 1–38 (1977)
76. Kirchner, M.: Fast and reliable resampling detection by spectral analysis of fixed linear predictor residue. In: Proceedings of 10th ACM Workshop Multimedia Security—MM&Sec'11 (2008)
77. Mahdian, B., Saic, S.: Blind authentication using periodic properties of interpolation. IEEE Trans. Inf. Forensics Secur. **3**, 529–538 (2008)
78. Li, S.P., Han, Z., Chen, Y.Z., Fu, B., Lu, C., Yao, X.: Resampling forgery detection in JPEG-compressed images. In: Proceedings of 2010 3rd International Congress on Image Signal Process CISP 2010, vol. 3, pp. 1166–1170 (2010)
79. Lien, C.-C., Shih, C.-L., Chou, C.-H.: Fast Forgery detection with the intrinsic resampling properties. J. Inf. Secur. **1**, 11–22 (2010)
80. Qian, R., Li, W., Yu, N., Hao, Z.: Image forensics with rotation-tolerant resampling detection. In: IEEE International Conference on Multimedia Expo Workshops ICMEW, pp. 61–66 (2012)
81. Birajdar, G.K., Mankar, V.H.: Blind method for rescaling detection and rescale factor estimation in digital images using periodic properties of interpolation. AEU Int. J. Electron. Commun. **68**, 644–652 (2014)
82. David, V., Fernando, P.: A Random Matrix Approach to the Forensic Analysis of Upscaled Images. IEEE Trans. Inf. Forensics, XX (2017)
83. Wang, R., Ping, X.J.: Detection of resampling based on singular value decomposition. In: Proceedings of Fifth International Conference on Image Graph, pp. 879–884 (2009)
84. Feng, X., Cox, I.J., Doërr, G.: Normalized energy density-based forensic detection of resampled images. IEEE Trans. Multimed. **14**, 536–545 (2012)
85. Hou, X.D., Zhang, T., Xiong, G., Zhang, Y., Ping, X.: Image resampling detection based on texture classification. Multimed. Tools Appl. **72**, 1681–1708 (2013)
86. Gloe, T., Ohme, R.: The dresden image database for benchmarking digital image forensics. In: ACM Symposium on Applied Computing, pp. 1584–1590

87. Qiao, T., Zhu, A., Retraint, F.: Exposing image resampling forgery by using linear parametric model. Multimed. Tools Appl. (2017)
88. Su, Y., Jin, X., Zhang, C., Chen, Y..: Hierarchical image resampling detection based on blind deconvolution q. J. Vis. Commun. Image Represent. 1–11 (2017)
89. Peng, A., Wu, Y., Kang, X.: Revealing traces of image resampling and resampling antiforensics. Adv. Multimed. (2017)
90. Bayar, B., Stamm, M.C.: On the robustness of constrained convolutional neural networks to JPEG post-compression for image resampling detection. In: IEEE International Conference on Acoustics Speech Signal Process, pp. 2152–2156 (2017)
91. Lin, X., Li, C., Hu, Y.: Exposing image forgery through the detection of contrast enhancement. In: International Conference on Image Process, pp. 4467–4471 (2013)
92. Stamm, M., Ray, K.J.: Blind forensics of contrast enhancement in digital images. In: Proceedings of International Conference on Image Process ICIP, pp. 3112–3115 (2008)
93. Cao, G., Zhao, Y., Ni, R.: Detection of image sharpening based on histogram aberration and ringing Artifacts. In: IEEE International Conference on Multimedia and Expo, pp. 1026–1029 (2009)
94. Cao, G., Zhao, Y., Ni, R., Kot, A.C.: Unsharp masking sharpening detection via overshoot artifacts analysis. IEEE Signal Process. Lett. **18**, 603–606 (2011)
95. Cao, G., Zhao, Y., Ni, R., Li, X.: Contrast enhancement-based forensics in digital images. IEEE Trans. Inf. Forensics Secur. **9**, 515–525 (2014)
96. Ding, F., Zhu, G., Yang, J., Xie, J., Shi, Y.Q.: Edge perpendicular binary coding for USM sharpening detection. IEEE Signal Process. Lett. **22**, 327–331 (2015)
97. Zhu, N., Deng, C., Gao, X.: Image sharpening detection based on multiresolution overshoot artifact analysis. Multimed. Tools Appl. (2016)
98. Hsu, Y.F., Chang, S.F.: Detecting image splicing using geometry invariants and camera characteristics consistency. In: International Conference on Multimedia and Expo, pp. 549–552 (2006)
99. Dong, J., Wang, W.: CASIA tampered image detection evaluation database
100. Dong, J., Wang, W.: CASIA2 tampered image detection evaluation (TIDE) database
101. Tralic, D., Zupancic, I., Grgic, S., Grgic, M.: CoMoFoD—New database for copy-move forgery detection
102. Amerini, I., Ballan, L., Caldelli, R., Bimbo, A. Del, Serra, G.: A SIFT-based forensic method for copy—move attack detection and transformation recovery. IEEE Trans. Inf. Forensics Secur. 1099–1110 (2011)
103. Amerini, I., Ballan, L., Caldelli, R., Del Bimbo, A., Del Tongo, L., Serra, G.: Copy-move forgery detection and localization by means of robust clustering with J-Linkage. Signal Process. Image Commun. 659–669 (2013)
104. Zandi, M., Mahmoudi-Aznaveh, A., Mansouri, A.: Adaptive matching for copy-move forgery detection. In: IEEE International Workshop on Information Forensics and Security, pp. 119–124 (2014)
105. Xie, D., Liang, L., Jin, L., Xu, J., Li, M.: A benchmark dataset for facial beauty perception. http://www.hcii-lab.net/data/SCUT-FBP
106. Bas, P., Filler, T., Pevný, T.: Break our steganographic system: the ins and outs of organizing BOSS. In: Proceedings of Information Hiding, Prague, Czech Repub, pp. 59–70 (2011)
107. Stich, M., Schaefer, G.: UCID—an uncompressed colour image database. In:Proceedings of SPIE, Storage Retrieval Methods and Application Multimedia, pp. 472–480 (2004)

# Comparative Study of Digital Forensic Tools

**Mayank Lovanshi and Pratosh Bansal**

## 1 Introduction

Digital data prevails all around us and plays a crucial role in any kind of investigation when cybercrime comes in a picture. Digital data comprises of binary representations and contains information in the form of text, images, audio, video, etc. In the present scenario, many cybercrime cases such as hacking, banking frauds, phishing, email spamming, etc., have emerged which are linked with digital data. Digital forensics is a new and demanding branch in the field of Computer Science [1]. Digital forensics is a scientific approach of preserving, acquiring, analyzing, extracting, and reporting of Digital evidences which come from the Digital sources like computer, mobile, camera, etc. It is categorized into various subbranches that are listed below as shown in Fig. 1.

- Computer Forensic
- Network Forensic
- Cyber Forensic
- Mobile Forensic
- Operating System Forensic
- Live forensic, etc.

There are many branches of forensic science present as discussed above, but we are working on desktop forensic, network forensic, and live network forensics.

1. **Desktop Forensic**: Desktop Forensic is a branch of digital forensic which is used for the extraction of digital evidence from the secondary memory. It deals with the recovery of the deleted files. Recovery is an important concept in cybercrime.

M. Lovanshi (✉) · P. Bansal
Department of Information Technology, IET, Devi Ahilya Vishwavidyalaya, Indore, MP, India
e-mail: lovanshi123mayank@gmail.com

P. Bansal
e-mail: pratosh@hotmail.com

© Springer Nature Singapore Pte Ltd. 2019
R. K. Shukla et al. (eds.), *Data, Engineering and Applications*,
https://doi.org/10.1007/978-981-13-6351-1_15

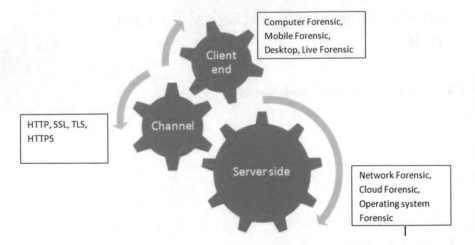

**Fig. 1** Digital forensic classifications

Here, computer is used as a target or as a source of digital crime. Desktop forensic is a type of digital forensic where we can determine the information from the hard disk, operating system. There are many software tools required for the recovery of the deleted file, i.e., Prodiscover basic, Cyber check suit, FTK analyzer, Recuva, Ease Us, etc. [2].

2. **Live forensics**: Live forensics is a branch of digital forensics which is used for the extraction of the digital evidence from the primary memory mainly focused on the RAM data. Here, RAM data like browsers information, cookies, registry, etc., are used as digital evidence in the live forensic case. It deals with the RAM dumping. Dumping of RAM means to extract information related to the RAM. There are many software tools present for extraction of the RAM data. Some tools are open source tools while some are licensed version tools like OSF Mount, Win-Lift, Belkasoft, Volatility Framework, etc. [2].

3. **Live network forensics**: Live network forensic tool is the branch of digital forensics. It deals with the live packet sniffing, packet spoofing, identification of the topology, etc. Here, mainly focused on the extraction of the digital evidence through the live network. In the live network, forensic packet information can be extracted. There are many live network forensic tools present like NMAP, wireshark, Ettercap, Nessus, etc. [2].

A number of authors suggested several scientific approaches for digital forensic investigation and defined some phases for the same which are shown in Fig. 2 and summarized as below

- **Identification**: In this phase, the evidence is identified from a digital source.
- **Collection**: Here, the evidence is seized, collected, or recorded from some digital source.

**Fig. 2** Digital forensic phases

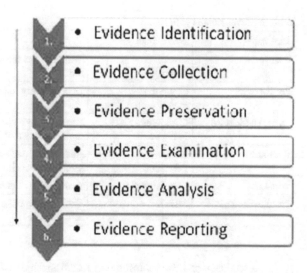

- **Preservation**: In this phase, digital evidence is preserved which helps in successful investigation, litigation, or incident response.
- **Examination**: Here, the evidence is examined or tested by the forensic expert which results in the correctness of evidence.
- **Analysis**: In this phase, digital forensic tools can be used and determine the relevant information from an image which has been seized during the preservation phase.
- **Reporting**: When an investigation is completed, then auditing and other meta information are reported under this phase [3].

Several digital forensics tools play a significant role in the process of digital forensics to carry out the investigation effectively. Some of the advantages of using tools are

- Used for producing digital evidences that are justifiable by the court of law.
- Helpful in proving the authenticity of evidences.
- Effective in reporting and documentation of evidences.
- Used for recovery of deleted data in various organizations.
- Helpful in research purposes in the domain.

In this paper, an effort has been made to compare different digital forensic tools and to provide a comparative study between them. This study determines the best tool on the basis of available parameters.

## 2 Literature Review

There are many authors who have given a comparative study of different forensic tools. Garber Lee explains the Encase tool which is a desktop forensic tool. The

**Table 1** Issues related with tools

| Tool name | Tool category | Description | Issues |
|---|---|---|---|
| Encase | Computer Forensic | Multipurpose forensic tool | Searching time and cost is too high |
| Prodiscover | Computer Forensic | Data recovery tool | Performance issue |
| Cyber check suite | Computer Forensic | Recovery and multipurpose tool | Low accuracy then other tool |
| Wire shark | Network Forensic | Analyzer of packet send on network | No intrusion detection system |
| Win-Lift | Live Forensic | Analyze RAM detail | Security issue |
| Mobile check | Mobile Forensic | Extract the mobile information | Incompatible with most of mobile |

author also described the features and functionalities of the tool. As per the author, Encase is a traditional tool whose results can be used in the court of law [3]. Abbas Cheddad explained a comparative study on cyber forensics tools such as Minitool, Hard Drive, and Pen drive Recovery tool based on the parameters like paper size and time, cost, tools availability, etc., and considered mapping as the biggest issue [4]. Nilakshi Jain tested digital forensic tools on different parameters in network, live, and desktop applications. The author also tested various other tools such as Clonezilla, Image USB, Etthercap, OSF Mount, etc. [5]. Kresimir Duretec worked on benchmarking of different forensic tools and also has used some data sets to verify the results. In his paper, he also suggested the reliability and security of digital evidences [6]. Ryan tested digital forensic tools such as Ease US, Encase, Ftk analyzer, etc., on the different parameters in the desktop forensic [7]. Abbas suggested [4] that comparative study of cyber forensic tools is present and gives the comparative study on Minitool, Hard Drive, Pen drive Recovery tool. In this paper, size and time is compared. Selection of the best tool is the biggest issue which was shown in this paper. Nilakshi Jain suggested [5] many Desktop Forensic tools tested on the different parameter for the Network, Live, and Desktop application. In this paper, traditional investigation process is used for the comparative study which will be not suitable for the court. Duretec et al. [6] suggested benchmarking of different forensic tools under which some data sets used to check the results on the different parameters. There are many tools issue extracted which will be shown in Table 1.

The review of the literature identified some problems which are as below

- Selection of tool—To select the best suitable tool based on the comparative study.
- Large computational time—It refers to huge computational time incurred while running some of the forensic tools.

# 3 Digital Forensic Tools

There are many forensic tools available in the category of freely available tools as well as the licensed version. Authors suggested the tools to identify evidences on the basis of digital forensic classification. Various types of digital forensic are discussed in Sect. 1 such as desktop forensic, live forensic, network forensic, etc., which are discussed below

## 3.1 Desktop Forensic

This is the computer forensic branch where the investigator focuses on the recovery of the secondary memory, i.e., hard disk of the system. Following are the tools to test desktop forensics applications:

a. **ProDiscover Basic**: The ProDiscover Basic is used to test the hard disk data. It is a forensic tool which is used to take legal action in court of law. It helps to collect data and imaging of that data and then to examine their recovery. It has very nice searching capability which allows data to be searched easily which is to be recovered. It allows data into a cluster view and content view [8].

b. **Encase**: Encase Digital Forensic tool is used to get, analyze, classify, recover, and reconstruct the evidence and test the digital evidence which is obtained from Digital Forensic investigation process. Encase is also a known traditional tool and its result is used into the court of law [8].

c. **Recuva**: Recuva is a data recovery tool program for Windows which is freely available in the market. It is used to recover permanently deleted files which are not overwritten. This tool can also be used to recover deleted files from USB, Hard Disk, and MP3Player [8].

d. **Cyber Check Suit**: This is also a cyber forensics tool for data recovery and analysis of digital evidence. Cyber Check needs imaging created by True Back image tool.

   The graphical user interface of the cyber check is very easy for a beginner. Cyber Check can also generate a report of the analysis findings by the investigative expert to submit in a court of law. The tool can report unallocated and disk slack area and gives options to do analysis based on file hashing and file's signatures [8].

e. **Autopsy**: Autopsy is an open source tool which is plugged-in with the Sleuth Kit collection. The GUI of this tool displays the results from the forensic searching of tool done by investigators to show an important part of the data. The development company provides the proper guidance of the tool [8].

## 3.2 Live Forensic

Live Forensic is a branch of Digital forensics where investigator focuses on the RAM feature extraction. Following are the tools to test live forensics applications:

a. **Win-Lift**: Win-Lift Analyzer is a forensic tool used for live analysis for analyzing RAM data collected by Win-Lift Imager. It results into forensic evidence and produces a full report. The analyzed information will be used for the proper reporting of the data. It results in different memory forensic object. It will extract the running files, log, open socket file, etc. [9].

b. **Belkasoft**: Belkasoft tool helps an investigator to acquire, search, analyze, and allocate digital evidence inside a computer. This tool is used to extract digital evidence from many sources by analyzing from various volatile memory data in term of evidence. This tool automatically exams the data source and put the RAM data for an expert to review, examine evidence data which has been analyzed by the tool [9].

c. **Magnet RAM**: Magnet RAM Capture is a tool which supports Windows systems including XP, Vista, 7, 8, and 10. It will extract the full live memory evidence and analyzes a volatile trace. This tool is used for memory analysis which is important for detection of malware and recovering valuable data. It extracts the running program and processes, active network connections, registries, password, keys, etc. These are just a few examples of the evidence that can be extracted from memory [9].

d. **OSF Mount**: OSF Mount is used to create RAM disk and this disk is mounted into RAM. This is used for the high-speed memory over the hard disk. This tool is useful in various applications like database application, cache files and browsers files. Another benefit is security, like data in the volatile disk will be automatically erased when a system is shutdown [9].

e. **Volatility Framework**: Volatility is open source tool for analyzing RAM evidence. This tool supports Linux, Windows OS. It is coded in Python and run On Windows, Linux system. It can examine RAM dump, crash dumps, virtual box dumps, etc. [9].

## 3.3 Network Forensic

Network forensic is a field of Digital Forensic where the investigator focuses on packets traveling in a network. Following are the tools to test network forensics applications:

a. **Wireshark**: Wireshark is an industry standard packet analyzer tool that analyzes the packets into a network. It is required in many projects that can be developed using the network. This tool is used to read and write the packet and capture file. This capture file will be analyzed by Wireshark tool. Many more tools are present but Wireshark is one of the best among them [2].

b. **Ettercap**: This tool is used to analyze the man in the middle attack and is a free and open source tool. This tool is used to analyze the packet tracing, security auditing, and computer network protocols. It is able to manage traffic on the network segment and capturing password [2].

c. **Nmap**: Nmap is a port scanner tool inside the network protocol which provides security features. It is used to discover host and provide services on the computer network. So it is used to map the services with network. Basically, Nmap (Network Mapper) is used to send a craft packet to target host and analyze the content and port of the target and source information [2].

d. **Nessus**: Nessus is a security auditing tool which is used to scan the network. It will scan a computer and create a warning about the vulnerability generated by malicious hackers. It can enter any computer which is connected with the network. This tool is testing different attacks into the computer which harm our computer system [2].

e. **Snort**: Snort is an open source tool which is used to detect network traffic, analyzes and packet logging on different IP logging system. It helps in protocol searching and packet tracing. Snort is used for providing QOS in a network. So this tool is capable of analyzing bulky data into a network [2].

# 4  Methodology

The methodology adopted by us focuses on the study of different digital forensic tools as discussed in the previous section. These tools describe various functionalities with respect to digital evidences. In our study, we follow a process as discussed in Fig. 3. The first step in the process involves selection of tools where we have chosen the digital forensic tools. The next step deals with the extraction of features which helps us in identifying the parameters. At last, we compare the tools on the basis of parameters.

The process helps us to determine whether the parameters are present in the respective tool or not. The parameters being considered are listed below

**Fig. 3** Flow chart of work

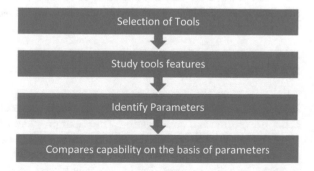

- Imaging: Imaging is a bit to bit copy of the hard drive. It takes every 0 and 1 from one hard drive to another.
- Hashing: Hashing uses the hash function to check the integrity of data used to verify the image of the drives.
- Recovery: Recovery is a process of getting back of data from the deleted drive. It is used to extract existing data.
- Acquire: It refers to identify the digital evidence inside the hard drive.
- Seizer: Seizer is used to preserve the hard drive by imaging.
- RAM Dumping: It refers to extract the RAM data such as cookies, browser history, registry information, etc.
- Live Log: It contains the logical files that are generated in live cases.
- Live Analysis: Live analysis helps to extract the RAM information from the dumping image of the RAM.
- Search: Search means to find some content into the image as well as the analyzed hard drive.
- Log: Log is a logical file that can be extracted from the analysis of the data.
- Reporting: Reporting means proper documentation generates after performing of the forensic tool.
- Packet Sniffing: It extracts information about the packets traveling in the network.
- Packet Analyzer: It is used to analyze the packet information such as IP, MAC, Firewall information etc.
- Packet Spoofing: Spoofing means hiding the information of the sender. Here, IP information is extracted.
- Protocol: It displays the rules followed by the tools.
- Open Port: It helps us to detect the open ports in IP connection that is required for application and servers.
- Topology: It refers to the arrangement of network representation.

## 5   Results

The comparative table contains the results of the different forensic tools based on the given parameters which are used to determine which tool is better in context to the parameters. Table 2 shows the comparative results for the tools.

## 6   Conclusion

In today's scenario, many digital forensics tools and techniques are used for cyber-crime prevention and investigation. The paper provides a comparative study between forensics application tools and a set of parameters. This approach is useful for forensics experts and investigators to select the best possible forensic tool based on their requirements. So, the investigation process can be carried out smoothly.

**Table 2** Comparative table result

| Tools | Parameter | | | | | | | | | | | | | | | | | | |
|---|---|---|---|---|---|---|---|---|---|---|---|---|---|---|---|---|---|---|---|
| | Digital forensic type | Tools availability | Imaging | Hashing | Recovery | Acquire | Seizer | RAM dumping | Live log | Live analysis | Search | Logs | Reporting | Packet sniffing | Packet analyzer | Packet spoofing | Protocol | Open port | Topology |
| Prodicover basic | Desktop | Trial | | ✓ | ✓ | | ✓ | – | – | – | ✓ | ✓ | ✓ | – | – | – | – | – | – |
| Encase | | Trial | ✓ | ✓ | ✓ | ✓ | ✓ | – | – | – | ✓ | – | ✓ | – | – | – | – | – | – |
| Recuva | | Free | – | – | ✓ | ✓ | ✓ | – | – | – | – | – | – | – | – | – | – | – | – |
| Cyber check suit | | License | ✓ | ✓ | ✓ | ✓ | ✓ | – | – | – | ✓ | – | ✓ | – | – | – | – | – | – |
| Autopsy | | Trial | ✓ | – | ✓ | ✓ | ✓ | – | – | – | – | – | – | – | – | – | – | – | – |
| Win-Lift | Live | License | ✓ | ✓ | – | ✓ | ✓ | ✓ | ✓ | ✓ | ✓ | ✓ | ✓ | – | – | – | – | – | – |
| Belkasoft | | Trial | ✓ | – | – | ✓ | ✓ | ✓ | ✓ | ✓ | – | – | – | – | – | – | – | – | – |
| Magnet RAM | | Trial | ✓ | ✓ | – | – | ✓ | ✓ | ✓ | ✓ | – | – | – | – | – | – | – | – | – |
| OSF mount | | Trial | ✓ | – | – | – | ✓ | ✓ | ✓ | ✓ | ✓ | ✓ | – | – | – | – | – | – | – |
| Volatility framework | | Trial | – | ✓ | – | – | ✓ | ✓ | ✓ | ✓ | – | ✓ | – | – | – | – | – | – | – |
| Wireshark | Network | Free | – | – | – | – | – | – | – | – | – | ✓ | – | ✓ | ✓ | ✓ | ✓ | | – |
| Ettercap | | Free | – | – | – | – | – | – | – | – | – | – | – | ✓ | ✓ | ✓ | ✓ | – | ✓ |
| Nmap | | Free | – | – | – | – | – | – | – | – | – | – | – | ✓ | ✓ | ✓ | ✓ | ✓ | ✓ |
| Nessus | | Free | – | – | – | – | – | – | – | – | – | – | – | ✓ | ✓ | ✓ | ✓ | ✓ | ✓ |
| Snort | | Free | – | – | – | – | – | – | – | – | – | – | – | ✓ | ✓ | ✓ | ✓ | ✓ | – |

After reviewed the comparative table we analyzed that some freely available tools are as useful as the license tool. Some tools are user-friendly in their GUI, some are better according to accuracy while some are used because of their lower data rate. This research is useful to show which tool is better on which condition and also concluding its usefulness of tool for stopping the digital crime.

## 7   Future Work

The future works include the mapping of digital forensic tools and enhancement of data accuracy, reliability, security, and other privacy measures by performing a comparative study on a large number of forensic tools.

In the next years, many other tools and many other features can also be used for future research. And provide the comparative result between that tool and parameter and we will try to be getting some CFTT research for testing the tools, also several other parameters may also include for comparison.

## References

1. Baryamureeba, V., Tushabe, F.: The enhanced digital forensic investigation process model. In: Proceedings of the 4th Annual Digital Forensic Research Workshop, Baltimore, MD, Citeseer (2004)
2. Mukkamala, S., Sung, A.H.: Identifying significant features for network forensic analysis using artificial intelligent techniques. Int. J. Digit. Evid. 1(4), 1–17 (2003)
3. Ani, L.: Cyber crime and national security: the role of the penal and procedural law. Law and Security in Nigeria, 200–202 (2011)
4. Cheddad, A., et al.: Digital image steganography: survey and analysis of current methods. Signal Process. 90(3), 727–752 (2010)
5. Jain, N., Kalbande, D.R.: A comparative study based digital forensic tool: complete automated tool. Int. J. Forensic Comput. Sci. (2014)
6. Duretec, K., Kulmukhametov, A., Rauber, A., Becker, C.: Benchmarks for digital preservation tools. In: Proceedings of IPRES 2015 (2015)
7. Hankins, R., Uehara, T., Liu, J.: A comparative study of forensic science and computer forensics. In: Third IEEE International Conference on Secure Software Integration and Reliability Improvement, 2009. SSIRI 2009. IEEE (2009)
8. Garber, L. Encase: a case study in computer-forensic technology. In: IEEE Computer Magazine January (2001)
9. System Administration Networking and Security Institute (SANS). Computer Forensics and Incident Response. https://www.sans.org/course/advancedcomputerforensic-analysis-incident-response (2015)

# A Systematic Survey on Mobile Forensic Tools Used for Forensic Analysis of Android-Based Social Networking Applications

Nirneeta Gupchup and Nishchol Mishra

## 1  Introduction

People today commonly use mobiles for their personal and organizational purposes. Mobiles are no longer used just for making calls, or some simple calculations, or clicking a quick photograph. They are used more often for web connectivity, chatting, social networking, e-mail, games, exchanging texts and messages, sending and receiving audio and video files, and so forth. The diverse features of mobile phones make it possible to increase the range of their uses in criminal activities as well.

**Social networking applications** provide a web page to a customer to create his own account on. From this account, the customer can generate his own list of people with whom he wants to share his thoughts, pictures, comments, personal information, and the like. He can also view the connected users' comments, pictures, thoughts, etc., within the system [1].

**Popularity**: According to Wikipedia, more than 7 billion people in the world use mobiles. China has the largest number of mobile phones in the world, 1.32 billion, while India is second with 1.12 billion mobile phones. The US with 327 million mobile phones comes third [2]. Social networking applications like Facebook, Twitter, MySpace, LinkedIn, Instagram, and WhatsApp have become widespread by means of mobile devices and vice versa. These kinds of social networking applications impact the lives of millions of people around the world.

Many famous politicians, actors, and sportsmen interact with their fans through the social media. They upload their pictures, their thoughts, and other data on social media networks. Their fans join them on the social media and frequently upload their own opinions, feelings, and other information. There are, for example, 29.8 M

N. Gupchup (✉) · N. Mishra
School of Information Technology, RGPV University, Bhopal, India
e-mail: nirneeta.gupchup@gmail.com

N. Mishra
e-mail: nishchol@rgtu.net

© Springer Nature Singapore Pte Ltd. 2019
R. K. Shukla et al. (eds.), *Data, Engineering and Applications*,
https://doi.org/10.1007/978-981-13-6351-1_16

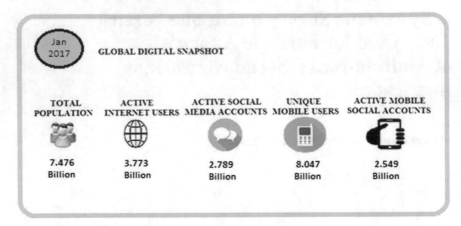

**Fig. 1** Active social media accounts worldwide in Jan 2017 [8]

followers of the Indian Prime Minister, Narendra Modi on Twitter [3]. The American President, Donald Trump has 29.3 M followers on Twitter [4]. Mr. Barack Obama has 88.0 M followers on Twitter [5]. Bollywood superstar Amitabh Bachchan has 26.7 M followers on the same microblogging site [6] and the famous cricketer Sachin Tendulkar has 15.9 M followers [7]. These statistics show the popularity of social networking applications (Fig. 1).

The research and consultancy organization, "We Are Social" announced in Jan 2017 that 2.789 billion people in the world had active social media accounts [8]. This huge figure shows the significance of social networking applications on Android and other systems.

**Uses**: There are some of the various uses of social networking apps:

- As already pointed out, social network services provide an account to a customer in which he can create a list of people with whom he wants to share his thoughts, pictures, comments, personal information, etc., and can also view the connected users' comments, pictures, thoughts, etc., within the system [1].
- Sometimes, missing people have been also found via social networking applications. For example, US airman Don Gibson found his long-lost son Craig and Don's wife Chrissie over Facebook after a lapse of 20 years [9]. In another similar incident, a California mother Holland reunited with his son Jonathan on Facebook after 15 years. Jonathan posted his childhood photo on Facebook and when Holland viewed it, she was able to reunite with her son [10].
- Social networking apps are now covering live debates. In America, at August 2015, the Republican presidential debate was conducted on Facebook and it was cohosted by Fox News [11].
- Several businessmen are using social networking to communicate with customers, for example, by forwarding coupons on Facebook and Twitter [12].

**Threats and vulnerabilities**: The diverse nature and popularity of the social media makes it highly vulnerable to attacks like phishing, stalking, child pornography, theft, social engineering attacks, etc. Cybercriminals can register themselves with fake identities and fake values with the intention to harm people or to carry out criminal activities.

A real legal case from the Sultanate of Oman involving a smartphone crime shows that a crime can be committed by using social networking applications on mobile. In this case, an organization (P) had received a user (Ali)'s complaint that his mobile phone had been hacked. And the numbers of his contact list were receiving text messages through WhatsApp. However, the user claims that he had not sent any messages from his mobile [13].

Some scams that have been perpetrated using social networking sites:

1. There was a link in Facebook which asked users to install WhatsApp on their PCs. When a user followed the link and did so, he had subscribed to a premium rate SMS service without his knowledge [14].
2. An attack on Twitter was carried out through direct messages, these DM appear to be come through a follower, and when we click on this messages link, it can either infect our twitter account or can steal account information [14].

**Need of mobile forensics**: When a case comes for investigation the police and investigators go through social media accounts to know about the victim and the suspect's general behavior. Social networking applications on mobiles are a profitable idea for forensic investigators. Electronic proof can be extracted from mobiles from the respective social media application. This gives information about user activities on the social media, as well as personal information. Following is the list of world's most popular 15 social networking applications:

1. Facebook,
2. YouTube,
3. Instagram,
4. Twitter,
5. Reddit,
6. Vine (In jan 2017 Vine became Vine Camera),
7. Pinterest,
8. Ask.fm,
9. Tumblr*,
10. Flickr,
11. Google+,
12. LinkedIn,
13. VK,
14. ClassMates, and
15. Meetup [15]

**Table 1** The worldwide device shipments by segment, 2016–2019 [41]

| Device type | 2016 | 2017 | 2018 | 2019 |
|---|---|---|---|---|
| Traditional PCs | 219 | 205 | 198 | 193 |
| Ultramobile premium | 49 | 61 | 74 | 85 |
| **PC market total** | **268** | **266** | **272** | **278** |
| Ultramobiles (basic and utility) | 168 | 165 | 166 | 166 |
| **Computing devices market** | **436** | **432** | **438** | **444** |
| Mobile phones | 1888 | 1893 | 1920 | 1937 |
| **Total** | **2324** | **2324** | **2357** | **2380** |

## Objective and Motivation

Forensic inspection of smartphones is difficult. Smartphones are constantly updating their contents, that is, they are in auto sync mode. The auto sync mode automatically updates mobile data which causes problems in extracting evidence. The reason of this analysis is to decide whether actions performed on social media by Android operating systems contain data stored in the internal memory of mobiles so that the amount, location, and time of data stored can be extracted.

The proliferation of smartphones with diverse features, the ubiquitous nature of mobile networks, and the extremely large numbers of users present problems for the investigator.

According to a survey conducted by Gartner in January 2017, the worldwide device shipments by segment were as follows in the period 2016–2019 (Millions of units).

In Table 1, a survey of worldwide device shipments by segment for four years (2016–2019) is shown. It is clear from the above table that the PC market is falling and the mobile phones market is growing rapidly. But unfortunately, mobile forensics is lagging behind as compared to computer forensics. There should be more research in mobile forensics so that the forensic process is equipped to recover all forensically significant data from mobile phones.

**Android Platform**: It is an open source mobile device platform. Android is based on the Linux 2.6 kernel. Open Handset Alliance is used to manage it. The first Android platform mobile phones were started in 2008, and today, it is the most popular operating system in the world. It has become the first operating system ever to ship one billion smartphones, globally. Today Android mobile's share is nearly 81.7% of the mobile market [16]. According to "THE VERGE" Android and iOS shared 99.6% of total mobile market in the last quarter of 2016 [16]. Following table describes different mobile operating system with their market share and unit volumes for 2 years (2015–2016).

Table 2 shows some mobile operating systems with their market share for 4Quarters of 2 years (2015–2016). It is clear from the above table that Android has the lion's share of the market (nearly 82% of the total mobile market). This shows the popularity of Android in the world.

**Table 2** In 4Q16 worldwide smartphone sales by operating system (thousands of units) [42]

| Operating system | 4Q16 units | 4Q16 market share (%) | 4Q15 units | 4Q15 market share (%) |
|---|---|---|---|---|
| Android | 352669.9 | 81.7 | 325394.4 | 80.7 |
| iOS | 77038.9 | 17.9 | 71525.9 | 17.7 |
| Windows phone | 1092.2 | 0.3 | 4395.0 | 1.1 |
| BlackBerry | 207.9 | 0.0 | 906.9 | 0.2 |
| Others OS | 530.4 | 0.1 | 887.3 | 0.2 |
| Total | 431539.3 | 100.0 | 403109.4 | 100.0 |

## 2 Main Text

### 2.1 Brief Literature Review

Mutawa et al. [17] researched applications such as Facebook, Twitter, and MySpace on smartphone: Android, iOS, and BlackBerry operating systems were analyzed. The result shows that the data stored in various social networking applications could be seen and evidence can be easily extracted from iOS and Android with the help of certain software and hardware. According to them, odin3 was used to root the tested Android mobile device so that they can upload the rootkit. After uploading the rootkit, social networking applications were installed. Then various user activities were performed on these applications. The results show that the Facebook artifacts contain files which have user's data, activity performed by the user including chat message, created albums, mailbox, etc. The Twitter artifact contains files which have the table of posted information like tweets, pictures, number of friends, users, and other useful data. The third social networking application MySpace artifact contains cookies, cache files because it is a web-based application.

Mathavan et al. [18] use commercial tools like WinHex, HxD for the purpose of analysis. They proved the integrity of the acquired image by calculating the hash value before and after taking the image of the Android mobile phone.

Mahajan et al. [19] researched WhatsApp and Viber instant messengers. They mention the forensics and challenges of Android that a forensics methodology should be chosen such that original data should be unchanged. Forensic examination of applications and their databases is tough if the mobile phone is locked or data is encrypted or deleted. They used Cellebrite UFED Classic Ultimate to extract the files and folders. They conduct an experiment on both rooted and non-rooted mobile phone.

Curran and Cakmak [20] mentioned that the mobile social media applications generally creates database files, log files, XML files, and PLIST files to store most of the private and evidentiary data. Logical images of Android mobile is created using XRY. And these files can be retrieved by using open source tools.

The research of Lessard and Kessle [21] shows that physical image of a smartphone gives more information than the logical image of the smartphone. They also explained the necessities of rooting of mobile. FTK (Forensic Toolkit) tool was used for the examination of memory image files.

Al-Hadidi and AlShidhani [13] mentioned in their research that Android mobile device has five layers: application layer, the application framework, libraries, Android run time, and Linux kernel. Java is used to write applications in applications layer. The device memory of mobile stores portable software applications like "WhatsApp" data and the portable device MicroSD card also contain data of software applications. In the same paper, they describe Oxygen forensic suite and UFED Physical Analyzer Cellebrite.

Iqbal et al. [22] researched left artifacts of ChatON Instant Messaging application on two devices: Samsung Galaxy Note running Android 4.1, and an iPhone running iOS. In their paper, they implemented a situation of predefined actions on both devices and acquired a picture of them(devices). Then they performed manual examination of the picture and recognized left data of ChatON Instant Messaging application. SQLite Database Browser, and plutil tools were used. They were successful to get the data with timestamp.

Kausar [23] researched the new directions in the area of smartphone forensic analysis. She compared different forensic techniques with their platforms. The result of her comprehensive analysis shows that there is no such generic forensic tool or technique which can perform forensic analysis of all types of smartphones.

In INFOSEC [24], it is stated that for forensic analysis of WhatsApp we need: Android mobile with USB data cable and mobile phone drivers, WhatsApp_Xtract_v2.0, file browser (manager) for Android, Python for windows. It is also clearly stated in it that all data is stored in WhatsApp in a SQLite database.

Yasin et al. [25] use 4 IM protocols namely AIM, Windows Live Messenger protocol, Google Talk, and Yahoo Messenger for investigation between the suspect and the correspondent of criminal network. Various conversations are performed between the suspect and the correspondent. Then they collected information left in memory and inspected by EnCasev6.14 to confirm that several IM protocols were used in a discussion.

Walnycky et al. [26] forensically analyzed 20 instant messaging applications on Android. Network traffic was also analyzed. The result shows that they were able to reconstruct data like passwords and screenshots, images, videos, audio sent, messages sent, and profile pictures.

Simao et al. [27] discussed Android platform that consists of the operating system, software development kit, and its applications. A sandbox concept is used by Android operating system so that applications cannot access those areas which are not explicitly allowed. SQLite database, free and open software, is also used by Android operating system.

Thakur [28] discussed WhatsApp database hardware and software acquisition. Hardware analysis is done by UFED Physical Analyzer and software analysis by Zena forensics. Android applications are written into Java programming language. SDK tool compiles the code into an .apk archive file. She concentrates her paper on two areas: acquisition and analysis of WhatsApp data from (1) nonvolatile memory and from (2) volatile memory.

Vidas et al. [29] in their paper mention that mobile forensic needs data collection from mobiles. For this purpose, numerous constraints and required qualities on the procedure are considered. Data should be correctly copied from mobile. The collection process is a multistep method which needs a collection recovery image. After obtaining it, it is flashed. After loading, of collection recovery image, the device is rebooted into recovery mode and connected with computer that has ADB (to verify that the device is connected).

Sahu [30] discusses the process of finding evidence. He extracted information via WhatsApp_Xtract package and then he installed python programming language environment.

Vinod [31] discussed forensic investigation in three main phases of digital forensics, namely, acquisition, analysis, and reporting. The analysis had two forms: evidence recovery and expert analysis. Android Debug Bridge (ADB) is used for pulling data. ADB is a free utility of Android SDK, and it is depending on the ADB pull command which copies parts of the file system to the forensic workstation. To retrieve required data from mobile the mobile is rooted.

## 2.2  Forensic Tool Kit

**Android Debug Bridge (ADB)**: This client–server program is a multipurpose command line tool. ADB can control Android gadget over USB from a computer for copying files, installing and uninstalling apps, and more. After installing ADB devices command we can communicate with an Android smartphone or tablet. An ADB with a USB connected device, USB debugging must be enabled [32].

**Open Source Android Forensics (OSAF)**: it is an open source forensic tool for creating a framework for Android which has a main focus on investigating malware within Android applications [33].

**XRY** provides physical and logical solutions. It provides investigators full right to use all the probable methods to recover information from a mobile device. We can recover vital data from mobile [34].

**Fastboot**: It is a small tool and can be used to reflash partitions on mobile. It is a command based tool. It comes with Android SDK (Software Development Kit). One can do recovery by using it. It is used to update firmware and doesn't require recovery mode. It is for Android mobiles [35].

**UFED Physical Analyzer**
Following things are possible with this tool:

- **Decoding**: Data carving from unallocated space, JTAG decoding, apps decoding, and image carving are possible.
- **Analysis**: Malware detection, project analytics, timeline, map view, file view, and SQLite database viewer are possible.

- **Reporting**: Report generation, export chat message in conversation format, and exporting of e-mail are possible.
- Hash verification is also possible [36].

**WhatsApp_Xtract**: It is a tool which is used for backup of WhatsApp social networking application messages. It is a freely available tool. It opens the msg-store.dbWhatsApp SQLite database and creates a report in which all contact with their conversation is mentioned [37].

**Oxygen Forensic**: It is a freely available forensic tool. It can access data from on Android, BlackBerry, iOS, Windows Phone, etc. It can import device backups and images. Password can be extracted by it. It is preferred in the investigation because it can recover most of the deleted data [38].

**MOBILedit Forensic**: Data that can be retrieved is call history, text messages, multimedia messages, files, calendars, note, reminders, application data such as Skype, Evernote, Dropbox, WhatsApp, Facebook, Gmail, etc. Phone-related information like IMEI number, operating systems, IMSI, ICCID, and location area information can also be retrieved by it. Password bypassing is possible. Deleted data can be retrieved and supports Android, BlackBerry, iPhone, Symbian, Windows phone, Windows Mobile, Bada, Meego, Chinese phones, and CDMA phones [39].

**Android Forensic Toolkit Comparison**
In Table 3, the comparison of mobile forensic tools is shown. From the above table, it is clear that:

**ADB**: This command based tool can analyze phone entries, messages, call logs, and application files. It can detect malware and extract password. Recovery of data and deleted data is also possible with it. This is important as forensic perception.

**OSAF**: This freely available tool can analyze phone entries, messages, call logs, and application files. It can detect malware. Recovery of data and deleted data is also possible with it. This is important as forensic perception.

**XRY**: This tool can analyze phone entries, messages, call logs, and application files. It can detect malware and extract password. Forensic examiners require recovery of data and deleted data which is also possible with it.

**Fastboot**: It has the same features as ADB.

- **UFED Physical Analyzer**: This tool extracts data quickly and it has same features like XRY. Three types of extraction methods are available in Cellebrite UFED:

1. Logical extraction,
2. File system extraction, and
3. Physical extraction.

1. **Logical Extraction**: In this extraction device vendor API is used. API gives right to use to apps to communicate with device operating system and permit forensic records for extraction.
   **Method**: Mobile is connected to UFED through a USB cable. After connection, UFED makes read-only API call to request for user data from mobile. Phone

**Table 3** Android forensic toolkit comparison

| Tools features | ADB | OSAF | XRY | Fastboot | UFED Physical Analyzer | WhatsApp_Xtract | Oxygen | MOBILedit |
|---|---|---|---|---|---|---|---|---|
| Phone entries | Yes | Yes | Yes | Yes | Yes | – | Yes | Yes |
| Messages | Yes | Yes | Yes | Yes | Yes | – | Yes | Yes |
| Call logs | Yes | Yes | Yes | Yes | Yes | – | Yes | Yes |
| Application files | Yes | Yes | Yes | Yes | Yes | – | Yes | Yes |
| Open access | – | Yes | – | – | – | Yes | Yes | Yes |
| Command-based | Yes | – | – | Yes | – | Yes | – | – |
| Data recovery | Yes | Yes | Yes | Yes | Yes | Yes | Yes | Yes |
| Recovery of deleted data | Yes | Yes | Yes | Yes | Yes | Yes | Yes | Yes |
| Password bypass-ing/password extracting | Yes | – | Yes | Yes | Yes | – | Yes | Yes |
| Malware detection | Yes | Yes | Yes | Yes | Yes | – | – | – |

replies to valid API requests and gives data such as chats, phonebook data, images, etc.

**Benefits**: Implementation is easy. This gives result in a readable format.

**Drawback**: Limited content can be extracted. Pictures are in a different folder hence cannot be extracted through it.

3. **Physical Extraction**: In this, a bit-to-bit copy of the mobile device's flash memory is created, which gives access to allocated and unallocated space of additional data layers [40].

2. **File System Extraction**: To access device partition this method is used. It has different set of built in protocols for different operating systems. In some cases, it is important to rely on phone backup like web history, e-mail headers.

**WhatsApp_Xtract**: This forensic tool forensically analyzes only WhatsApp social networking application data. It is a command based tool.

**Oxygen**: This freely available tool analyzes phone entries, messages, call logs, and application files. Forensic important data can be recovered by this tool. Forensic examiners can extract password using this tool.

**MOBILedit**: This tool installs a small application on the mobile phone to pull the data. This has the same feature like oxygen.

## 3 Conclusion and Future Work

Several forensic toolkits: ADB, OSAF, XRY, Fastboot, UFED Physical Analyzer, Whatsapp_Xtract, Oxygen, and MOBILedit are analyzed in this paper. UFED and XRY forensic tools are able to extract required data. These tools are analyzed on the features like phone entries, messages, call logs, application files, open access, command based, data recovery, recovery of deleted data, password extracting, and malware detection. In this paper, some of world's famous social networking applications are analyzed on Android operating system. In future, other mobile operating systems like iOS, BlackBerry, and Windows will be analyzed for forensic analysis and extracting data.

## References

1. Social networking service. https://en.m.wikipedia.org/wiki/Social_networking_service
2. List of countries by number of mobile phones in use. https://en.m.wikipedia.org/wiki/list_of_countries_by_number_of_mobile_phones_in_use
3. Narendra Modi. https://twittercounter.com/narendramodi
4. Donald Trump. https://twittercounter.com/realDonaldTrump
5. Barack Obama. https://twittercounter.com/BarackObama
6. Amitabh Bachchan. https://twittercounter.com/SrBachchan
7. Sachin Tendulkar. https://twittercounter.com/Sachin_rt
8. Global social media research summery 2017. http://www.smartinsights.com/social-media-marketing/social-media-strategy/new-global-social-media-research/
9. Facebook fairytale: dad traces long-lost son online after 20 years… and falls in love with mum again too. http://www.mirror.co.uk/news/real-life-stories/dad-traces-long-lost-son-on-facebook-1150447
10. Facebook photo helps reunites mother with son who was kidnapped 15 years ago. www.dailymail.co.uk/news/article-3148543/Facebook-photo-helps-reunites-mother-son-kidnapped-15-years-ago.html
11. How Facebook, Twitter and Google Trends covered the Republican debate. http://www.livemint.com/Consumer/nGGaJtZECOPylJyBzheNrN/How-Facebook-Twitter-and-Google-Trends-covered-the-Republic.html
12. The importance of Social Media in our daily life. http://kaklaw-ccc-2012.blogspot.in/
13. Al-Hadidi, M., AlShidhani, A.: Smartphone forensic analysis: a case study. Int. J. Comput. Electr. Eng. 576–580 (2013)
14. Scams on social networks that will surprise you. http://www.pandasecurity.com/mediacenter/social-media/scams-social-networks-will-surprise/
15. Top 15 most popular social networking sites (and 10 Apps!). https://www.dreamgrow.com/top-15-most-popular-social-networking-sites/
16. Percent of new smartphones run Android or iOS. https://www.theverge.com/2017/2/16/14634656/android-ios-market-share-blackberry-2016
17. Al Mutawa, N., Baggili, I., Marrington, A.: Digit. Investig. S24–S33. Journal homepage www.elsevier.com/locate/diin (2012)
18. Mathavan, T., Nagoor Meeran, A.R.: Acquisition and analysis of artifacts from instant messenger on android device. Int. J. Eng. Res. & Technol. (IJERT), 1210–1212 (2014)
19. Mahajan, A., Dahiya, M.S., Sanghvi, H.P.: Forensic analysis of instant messenger applications on android devices. Int. J. Comput. Appl. 38–44 (2013)

20. Curran, K., Cakmak, A.Y.: Social media forensics on mobile devices (2015)
21. Lessard J., Kessler, G.C.: Android forensics: simplifying cell phone examinations. Small Scale Digit. Device Forensics J. (2010)
22. Iqbal, A., Marrington, A., Baggili, I.: Forensic artifacts of the ChatON instant messaging application. In: 8th International workshop on SADFE (2013)
23. Kausar, F.: New research directions in the area of smart phone forensic analysis. Int. J. Comput. Netw. & Commun. (IJCNC), 99–106 (2014)
24. INFOSEC. http://resources.infosecinstitute.com/android-whatsapp-chat-forensic-analysis/
25. Yasin, M., Kausar, F., Aleisa, E., Kim, J.: Correlating messages from multiple IM networks to identify digital forensic artifacts. Electron. Commer. Res. **14**, 369–387 (2014)
26. Walnycky, D., Baggili, I., Marrington, A., Moore, J., Breitinger, F.: Network and device forensic analysis of Android social-messaging applications. Digit. Investig. 77–84 (2015)
27. Simao, A.M. de L., Sicoli, F.C., de Melo, L.P., de Sousa Junior, R.T.: Acquisition of digital evidence in android smartphones. In: 9th Australian Digital Forensics Conference, Edith Cowan University, Perth, Western Australia (2011)
28. Thakur, N.S.: Forensic analysis of WhatsApp on android smartphones. University of New Orleans ScholarWorks@UNO (2013)
29. Vidas, T., Zhang, C., Christin, N.: Toward a general collection methodology for Android devices. Digit. Investig. **8**, S14–eS24 (2011)
30. Sahu, S.: An analysis of WhatsApp forensics in android smartphones. Int. J. Eng. Res. 349–350 (2014)
31. Vinod, J.: Forensic analysis of WhatsApp artifacts on android phones. Int. J. Adv. Res. Trends Eng. Technol. (IJARTET), 89–94 (2015)
32. Android debug bridge. http://developer.android.com/tools/help/adb.html
33. Open source Android forensics. http://osaf-community.org/
34. XRY mobile forensics. http://www.secureindia.in/?page_id=1128
35. Doc: fastboot intro. http://wiki.cyanogenmod.org/w/Doc:_fastboot_intro
36. UFED physical analyzer features list. http://www.cellebrite.com/Pages/ufed-Physical-analyzer
37. hotoloti/whatsapp_xtract.py. https://github.com/knomoseikei/hotoloti/blob/master/whatsapp_xtract.py
38. Oxygen forensics. http://www.oxygen-forensic.com/en/
39. Software used by millions for phone content management, transfer and investigation. http://www.mobiledit.com/
40. Explaining Cellebrite UFED data extraction processes. http://smarterforensics.com/wp-content/uploads/2014/06/Explaining-Cellebrite-UFED-Data-Extraction-Processes-final.pdf
41. Gartner forecasts flat worldwide device shipments until 2018. http://www.gartner.com/newsroom/id/3560517
42. Gartner says worldwide sales of smartphones grew 7 percent in the fourth quarter of 2016. http://www.gartner.com/newsroom/id/3609817

**Nirneeta Gupchup** Scholar of School of Information Technology, RGPV University, Bhopal, India.

**Nishchol Mishra** Ph.D., in Computer Science and Engineering, Assistant Professor in School of Information Technology, RGPV University, Bhopal, India.
**Research Area** Multimedia data mining and social media analytics.

# Enhanced and Secure Acknowledgement IDS in Mobile Ad Hoc Network by Hybrid Cryptography Technique

Aumreesh Kumar Saxena, Piyush Shukla and Sitesh Kumar Sinha

## 1 Introduction

MANET involves remote compact center points that shape a fleeting framework without the guide settled structure or central association [1]. Center points can confer particularly to various centers inside their transmission go. Center points outside the transmission extend are passed on by methods for transitional center points with the true objective that it shapes a multi-hop circumstance [1]. In multi-skip transmission, a package is sent beginning with one center then onto the following, until it accomplishes the objective with the help of using guiding tradition. For honest to goodness, working of the framework-coordinated effort between center points is required [1, 2]. Here, investment suggested playing out the framework limits out and out by centers for the preferred standpoint of various center points. In any case, since the open establishment and flexibility of center points, noncooperation may happen which can to a great degree degrades the execution of framework [1, 2]. MANET is weak against various sorts of strikes because of the open establishment, dynamic framework topology, nonattendance of central association and compelled battery-based essentialness of compact center points [3]. These ambushes can be named Denial of Service Attack, Impersonation, Eavesdropping Routing strikes, and Black crevice strike, Gray-opening Attack, Man-in-the-inside Attack, Jamming, Replay Attack, and Wormhole Attack [3]. A couple arranges had been proposed as of now that solely pointed on area and abhorrence of outside attacks [2, 3]. However,

A. K. Saxena (✉) · S. K. Sinha
AISECT University, Bhopal, India
e-mail: aumreesh@gmail.com

S. K. Sinha
e-mail: siteshkumarsinha@gmail.com

P. Shukla
UIT RGPV, Bhopal, India
e-mail: phdpwd@gmail.com

© Springer Nature Singapore Pte Ltd. 2019
R. K. Shukla et al. (eds.), *Data, Engineering and Applications*,
https://doi.org/10.1007/978-981-13-6351-1_17

an expansive part of these arrangements twists up perceptibly futile when the poi-
sonous center points starting at now entered the framework or a couple of centers in
the framework are exchanged off by attacker [3]. Such attacks are more perilous as
these are begun from inside the framework and because of this the primary Defense
line of framework winds up clearly unfit. Since inward attacks are performed by
taking an intrigue noxious center points which bear on quite a while before they are
exchanged off along these lines it ends up being extraordinarily difficult to recognize
[4]. MANET is prepared for making a self-planning and self-keeping up framework
without the help of a concentrated establishment, which is as often as possible infeasible in essential mission applications like military conflict or emergency recovery [4].
Unimportant setup and quick course of action make MANET arranged to be used as
a piece of emergency conditions where an establishment is blocked off or unfeasible
to present in circumstances like typical or human-incited cataclysms, military con-
flicts, and restorative emergency conditions [5]. Securing remote ad hoc framework
is astoundingly trying the issue.

## 2 Intrusion Detection in MANET

Various IDS have been presented in standard wired frameworks, where all developments must experience switches or entryways. IDS can be added to and executed in
these devices easily on the other hand; MANET works do not have such devices [6].
Moreover, the medium is absolutely open, so both genuine and harmful customers
can get to it [5, 6]. In addition, there is no sure division among average and remarkable activities in a convenient circumstance. Since centers can move subjectively,
false directing information could be from an exchanged off-center point or a center
that has old information. The present IDS procedures on wired frameworks cannot be
associated clearly to MANETs [6]. Various IDS projected to ensemble the attribute
of MANETs.

## 3 Watchdog

The essential of the watchdog part is to improve the throughput of the framework
with the closeness of threatening centers. Watchdog fill in as intrusion revelation
for MANET and responsible for perceiving noxious center unfortunate behavior in
the framework [6]. Protect watchdog perceives harmful center insidious exercises
by unpredictably tuning into its next ricochet's transmission [7]. In case a Watchdog
center point gets that, its next center point fails to forward the package inside a
predefined day and age, it fabricates its mistake counter [7]. At whatever point a
center's failure counter outperforms a predefined edge, the Watchdog center reports
it as getting boisterous [7, 8]. Meanwhile, protect watchdog keeping up a pad of
starting late sent packages and differentiating each got allocate the bundle in the

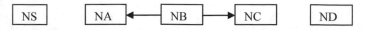

**Fig. 1** Watchdog concept

support [7, 8]. A data distribution cleared from the pad when the protect watchdog gets a comparable package is being sent by the accompanying bob center point over the medium. If a data distribution in the pad is for a truly long time, the watchdog plot accuses the accompanying hop neighbor to be escaping hand (Fig. 1).

Exactly when center point NB propels a package from center point NS toward center point ND through center point NC, center point NA cannot transmit the separation to center point NC, in any case, it can tune in on center point NB's traffic. Center point NA can get center point NB's transmission and can affirm that center point NB has tried to pass the package to center point NC. The solid line addresses the arranged course of the package sent by center point NB to center point NC, while the dashed line demonstrates that is inside transmission extent of center point NB and can get the bundle trade. The way rater technique empowers center points to keep up a key separation from the use of the raising hell center points in any future course selections. The directing information can be passed with the message [8]. The Watchdog plot fails to distinguish harmful wicked exercises with the closeness of the going with: (1) unverifiable accidents; (2) authority impacts; (3) compelled transmission control; (4) false terrible lead report; (5) interest; and (6) fragmentary dropping.

## 4 AACK

**AACK**: Another arrangement called AACK. Like TWOACK, AACK is a certification-based framework layer plot which can be considered as a blend of an arrangement called TACK (vague to TWOACK) and a conclusion to end insistence scheme called ACKnowledge (ACK) [8, 9]. Stood out from TWOACK, AACK in a general sense lessened framework overhead while still prepared for keeping up or despite outflanking a comparative framework throughput [9]. Enhance Adaptive Acknowledgment (EAACK) is planned to deal with two of the six deficiencies of Watchdog arrange, specifically, false fiendishness and recipient crash [8, 9].

## 5 TWOACK

**TWOACK**: It deals with the issue of the beneficiary crash and power hindrance of protecting watchdog. In this arrangement, a certification of every data allocates each there centers along transmission way [9]. If ACK is not gotten inside predefined time, interchange centers are stamped malicious. TWOACK tackles directing tradi-

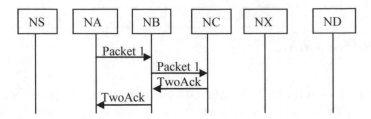

**Fig. 2** Two acknowledgement

tions, for instance, Dynamic Source Routing (DSR) [9]. The shortcomings are (1) Limited battery control (2) Network overhead. Figure 2 is showing the working of two acknowledgement schemes.

## 6 Related Work

In [9] Hybrid cryptography method is showed and which is decreased by controlling overhead by recognizing the malicious way. Using shared key source center point and objective center point confirm to trade data Packet [9]. Cream cryptography methodology uses AES and RSA Public Key Pair to send data from source to objective [9]. AES figuring is used for scrambling content which is unscrambled by same key [9]. It is as a kind of symmetric piece figure. It is related with the TA to ensure that symmetric keys are developed between genuine center points [9]. The data packages sent from the source center to Trust authority (TA) center and to the objective. Bundle drop is diminished by completing the puzzle key created for each flexible center point in the framework [9]. In [10] it is shown the system managing three issues of the watch pooch—that are confined transmission control, recipient crash, and false inconvenience making. Here in the presented system DSR tradition is used and to keep up a vital separation from the created confirmation and it uses cryptographic computation that is RSA [10]. In [11] secure check procedure is proposed which is cream cryptographic frameworks and exceptionally expected for MANETs to diminish the framework overhead in flexible center points. It develops and ensures the higher terrible lead reports to extend the framework execution while perceiving the groups in EAACK plot [11]. In [11] DSA and RSA, propelled check arrange is executed. EAACK is a certification-based IDS each one of the three areas of EAACK, to be particular ACK, S-ACK, and MRA, are attestation-based acknowledgment arranges [11]. In [12] cream strategy of RSA and AES is used, to make the system more secure as RSA figuring is used to talk with the gatherer through session key and AES estimation is used to scramble this session key which makes the key more secure along these lines redesigning the security level. In [13] it is displayed a structure which can perceive and what's more to keep the toxic strikes. The structure is named as Enhanced Adaptive ACKnowledgment (EAACK) [13]. This structure has a powerful strike control, which is one of the basic conditions to guarantee the data security [13]. Once the activator describes the keys to the centers, the need will be made therefore. What's more, each record would be secured at the activator database [13]. In [14] dangerous center point acknowledgment arranges is presented. Uncommon mode in the presented plan will recognize the threatening center point [14]. The malevolent center area handle is taken after just-in harmful circumstance. So if there is no malignant center point in the framework, by then the strategy goes about as whatever other coordinating tradition, e.g., AODV [14]. An additional mode in the proposed plan may grow the overhead, however if we are recognizing the genuine malignant center then the overhead is sufficient [14]. What's more, besides by making use of Elliptic curve cryptography with Digital stamp estimation will guarantee our framework. The Elliptic Curve Digital Signature count (ECDSA) takes lesser key size than RSA/DSA estimations for the same level of security [14].

# 7 Issues

Enhanced and Secured Adaptive Acknowledgment (ESAACK) is expected to deal with some weaknesses of Watchdog arrange, particularly, false rambunctiousness and authority affect.

## 7.1 Receiver Collisions

Center point NA sends Packet P1 to center point NB, further center point NB sends this bundle to center point NC; meanwhile, center point NX is sending Packet 2 to center point NC, due to which, packet 1 and packet 2 dropped and did not reach at center point NC, also message from center point NB is sent to center point NA of packet failure see Fig. 3.

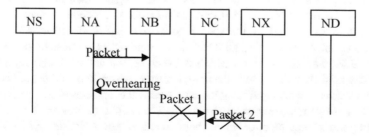

**Fig. 3** Receiver collision

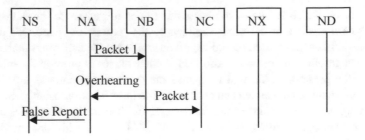

**Fig. 4** False misbehavior report

## 7.2 False Misbehavior Report

Center point NB sent Packet P1 to center point NC, due to some problem packet it was not received by center point NC but a false report is sent by center point NA to center point NS of successful transmission of the packet due to open medium and remote conveyance of regular MANETs resulting in misbehavior report see Fig. 4.

## 8 Proposed Work

ACK is usually called an end-to-end acknowledgment procedure which is a piece of EAACK procedure and an important aspect of ACK is to decrease the overhead in the network. For example, if node A1 is transmit a packet P1 to another node A2, and if each intermediate node are supportive and effectively received the request in the node A2. Then Node A2 will generate ACK to node A1. **Secure Acknowledgment (S-ACK)** principle work where three or more than three successive nodes in a network to identify misbehaving of the node. For every three successive nodes in the path, the third node compulsorily gives reply as S-ACK packet to the first node. The motive of S-ACK is the identification of misbehaving node during receiver collision and limited transmission power. To determine the flaws of watchdog like false misbehavior report we used **Misbehavior Report Authentication (MRA)** procedure. Here another or different path is checked by the source node, if it is found and data reached at destination node by using this path which is summarized as the false report. EAACK include ACK, SACK and MRA which is based on the acknowledgment concept. To identify misbehaviors node in the network, they used acknowledgement concept. Thus, it is tremendously significant to make sure that each acknowledgment packets in EAACK are genuine and uncontaminated. The motive of presented work is also achieved genuine and uncontaminated acknowledgement packet in ESAACK.

## 8.1 Proposed Architecture

The proposed approach presents Intrusion Detection System in MANETs. In the framework, first task is to make center points. Once the center points are made, they are associated and their accessibility is checked. The proposed system is for MANET. A cross-breed tradition is exhibited named as gathered coordinating tradition. It isolates an aggregate framework into different gathering. Intra-assemble coordinating are done by a proactive tradition which diminishes the delay in the correspondence of center points inside the group. The proposed cross-breed tradition is a mix of RSA and BE approach which diminishes the framework overhead. For this, the approach has lessened malignant center point through which security level is overhauled.

### 8.1.1 Network Formation

Keeping in mind to build the network, we are proceeding to embrace the accompanying execution measurements.

### 8.1.2 Cluster Routing Protocol (CRP)

CRP is the blend of a proactive and responsive protocol. There is a diminishment in the deferral of correspondence between center points in a framework if intra-gather coordinating is given by a proactive protocol. There is a diminishment in information exchange limit with respect to information in the middle of gathering coordinating by using a responsive protocol. CRP uses the framework called as periphery cast assurance tradition, which manages the development between dissimilar clusters. It comes into the picture when there is no periphery cast assurance tradition used to spread the responsive route inquire. If a center point has no route to its destination, it is used to spread the responsive route inquire.

## 8.2 Hybrid Cryptography Techniques

In the present structure ESAACK, there is a huge amount of framework overhead when there is an extension in a malevolent center point in the framework. These malignant center points enlarge the fake insistence, which realizes framework overhead. The structure uses RSA and BF estimations. In this proposed structure, we use the BF symmetric computation to scramble the plaintext and uses RSA used to encode the symmetric key of BF. For a session, individual key will produced.

# 9 Results

Java platform is used during experiments in this work. We have prepared computer network architecture in a computer lab where 15 wireless nodes connected with each other through the server node. During transmission of the packet between two nodes (source, destination), we choose nodes randomly. Sizes of the transmitted packet were 512 bytes and transmission rate were 3 packet/second. Every node in the network maintaining a buffer which can manage 80 packets during sending and receiving if routes are not available at current time. During experiments, if an intrusion is found, an alarm will generate in the form of dialog box which will display on nodes screen. We have evaluated the inside performance of the node. During experiments for the evaluation of the proposed method, we choose the following parameters:

1. **Packet Delivery Ratio (PDR)**: PDR can be described as the ratio of data packets received through destinations to those produced by the sources [15, 16].
2. **Delay**: Another important attribute of performance is the network delay. Delay can be explained as a total time taken by a bit of data to reach from source–to–destination node in the network [15, 16].
3. **Routing Overhead**: ratio of routing-related transmissions [15, 16].
4. **Packet Loss**: Packet loss is another important attribute in performance evaluation; it is defined as a total number of packet loss during transmission from source-to-destination node in the network [15, 16].

Here, the performance of the presented EAACK is examined and final result is concluded. Table 1 is showing the number of center point in the framework and corresponding delay. Graph 1 encircled from Table 1 and it demonstrates that the ESAACK structure gives the slightest Delay. The deferral is constrained since the package is passed on in slightest time. Table 2 is demonstrating the bundle lose on different parcel estimate in the system. Graph 2 framed from Table 2 and it shows that in the proposed framework parcel misfortune is diminished. As we realize that bundle misfortune is one reason for vitality squander moreover.

Table 3 is exhibiting the package movement extent on various package sizes in the framework. Graph 3 molded from Table 3 and we can express that bundle movement extent rate is high. On account of this component, a most extraordinary number of packages are passed on. Table 4 is showing the coordinating overhead on various package sizes in the framework. Graph 4 molded from Table 4 and it is showing that the proposed system diminishes the coordinating overhead.

**Table 1** Nodes versus average delay

| No. of nodes | ESAACK IDS average delay |
|---|---|
| 4 | 0.0245 |
| 6 | 0.035 |
| 8 | 0.0415 |
| 12 | 0.0389 |

**Graph 1** Node versus average delay

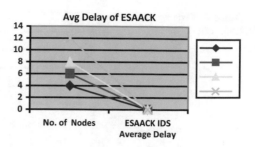

**Table 2** Size of packet versus packet lose

| No. of nodes | ESAACK IDS packet loss |
|---|---|
| 4 | 2 |
| 6 | 2 |
| 8 | 3 |
| 12 | 5 |

**Graph 2** Packets size versus loss

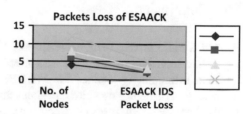

**Table 3** Node versus execution time PDR

| Nodes | ESAACK IDS PDR |
|---|---|
| 4 | 2.351 |
| 6 | 2.451 |
| 8 | 3.258 |
| 12 | 3.387 |

**Graph 3** Packet size versus PDR

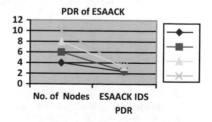

**Table 4** Node versus routing overhead

| Nodes | ESAACK IDS overhead |
|---|---|
| 4 | 2.351 |
| 6 | 2.451 |
| 8 | 3.258 |
| 12 | 3.387 |

**Graph 4** Conn versus routing overhead

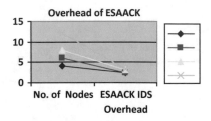

## 10  Conclusion

ESAACK build MANETs highly sheltered. The genuine risks like false terrible lead report and mold assertion can be recognized by using this arrangement. ESAACK tradition astoundingly planned for MANETs and examined it against other very distinctive parts in different circumstances through generations. Coming to the final result where we notified positive execution against existing arrangement, for instance, watchdog, TWOACK. The development in noxious centers in the structure achieves framework overhead and different fake certifications made. Thus, to diminish the framework overhead and fake insistence, we have used new approach with cryptography. In this methodology cross-breed, Clustered tradition reduces the threatening center, which decreases orchestrate overhead. Presented method beats the negative aspect of PKI (Public key establishment) and moreover improved MANET security. Execution of projected work is completed with the help of cryptographic methodology, (for instance, both BF and RSA plots in the amusement).

## References

1. Soni, M., Ahirwa, M., Agrawal, S.: A survey on intrusion detection techniques in MANET. In: International Conference on Computational Intelligence and Communication Networks (CICN), India, pp. 1027–1032 (2015)
2. Banerjee, S., Nandi, R., Dey, R., Saha, H.N.: A review on different intrusion detection systems for MANET and its vulnerabilities. In: International Conference and Workshop on Computing and Communication (IEMCON), India, pp. 1–7 (2015)
3. Elboukhari, M., Azizi, M., Azizi, A.: Intrusion detection Systems in mobile ad hoc networks: a survey. In: 5th Workshop on Codes, Cryptography and Communication Systems (WCCCS), Morocco, pp. 136–141, November 2014

4. Nemade, D., Bhole, A.T.: Performance evaluation of EAACK IDS using AODV and DSR routing protocols in MANET. In: International Conference on Emerging Research in Electronics, Computer Science and Technology (ICERECT), India, pp. 126–131 (2015)
5. Indira, N.: Establishing a secure routing in MANET using a hybrid intrusion detection system. In: Sixth International Conference on Advanced Computing (ICoAC), India, pp. 260–263 (2014)
6. Sakila Annarasi, R., Sivanesh, S.: A secure intrusion detection system for MANETs. In: IEEE International Conference on Advanced Communications, Control and Computing Technologies, India, pp. 1174–1178 (2014)
7. Sandhiya, D., Sangeetha, K., Latha, R.S.: Adaptive acknowledgement technique with key exchange mechanism for MANET. In: International Conference on Electronics and Communication Systems (ICECS), India, pp. 1–5 (2014)
8. Shakshuki, E.M., Kang, N., Sheltami, T.R.: EAACK—a secure intrusion detection system for MANETs. IEEE Trans. Industr. Electron. **60**(3), 1089–1098 (2013)
9. Awatade, S., Joshi, S.: Improved EAACK: develop secure intrusion detection system for MANETs using hybrid cryptography. In: International Conference on Computing Communication Control and automation (ICCUBEA), India, pp. 1–4 (2016)
10. Kazi, S.B., Adhoni, M.A.: Secure IDS to detect malevolent node in MANETs. In: International Conference on Electrical, Electronics, and Optimization Techniques (ICEEOT), India, pp. 1363–1368 (2016)
11. Gowthaman, G., Komarasamy, G.: A study on secure intrusion detection system in wireless MANETs to increase the performance of Eaack. In: IEEE International Conference on Electrical, Computer and Communication Technologies (ICECCT), India, pp. 1–5 (2015)
12. Patil, T., Joshi, B.: Improved acknowledgement intrusion detection system in MANETs using hybrid cryptographic technique. In: International Conference on Applied and Theoretical Computing and Communication Technology (iCATccT), India, pp. 636–641 (2015)
13. Joshi, P., Nande, P., Pawar, A., Shinde, P., Umbare, R.: EAACK—a secure intrusion detection and prevention system for MANETs. In: International Conference on Pervasive Computing (ICPC), India, pp. 1–6 (2015)
14. Patil, A., Marathe, N., Padiya, P.: Improved EAACK scheme for detection and isolation of a malicious node in MANET. In: International Conference on Applied and Theoretical Computing and Communication Technology (iCATccT), India, pp. 529–533 (2015)
15. Shakshuki, E.M., Kang, N., Sheltami, T.R.: EAACK—a secure intrusion detection system for MANETs. IEEE Trans. Industr. Electron. **60**(3) (2013)
16. Sanjith, S., Padmadas, M., Krishnan, N.: EAACK—based intrusion detection and prevention for MANETs using ECC Approach. Int. J. Emerg. Trends Technol. Comput. Sci. (IJETTCS) **2**(4) (2013)

# Formal Verification of Causal Order-Based Load Distribution Mechanism Using Event-B

Pooja Yadav, Raghuraj Suryavanshi, Arun Kumar Singh and Divakar Yadav

## 1 Introduction

Distributed systems are very complex to understand and develop. There is a need to formally verify and ensure the correctness of distributed systems and algorithms. During last few years, the research in the field of formal methods has done significant work in the development of describing and analysing the complex systems in formal languages [1, 2]. In distributed systems, formal methods take an important role for ensuring the correctness of several protocols and algorithms. Formal verification is done either through model checking or theorem proving [3, 4]. Model checking is a model-oriented approach, which verifies the correctness of system automatically by traversing every possible execution path. It is expressed in terms of finite state automata that describe all possible transitions states. In this technique, for a given problem, a formal model describing its behavioural properties is developed in formal language. All the properties of models are verified. In order to verify all possible execution paths, it is required that the model must be finite. The problem may also appear when the model which is finite has considerable size. Theorem proving is the act of generating a mathematical proof for a mathematical statement to be true

P. Yadav
Abdul Kalam Technical University, Lucknow 226031, India
e-mail: poojayadav255@gmail.com

R. Suryavanshi (✉)
Pranveer Singh Institute of Technology, Kanpur 209305, India
e-mail: raghuraj_singh09@yahoo.co.in

A. K. Singh
Rajkiya Engineering College, Kannauj 209732, India
e-mail: aksingh_uptu@rediffmail.com

D. Yadav
Institute of Engineering and Technology, Lucknow 226021, India
e-mail: divakar_yadav@rediffmail.com

© Springer Nature Singapore Pte Ltd. 2019
R. K. Shukla et al. (eds.), *Data, Engineering and Applications*,
https://doi.org/10.1007/978-981-13-6351-1_18

229

[3, 5, 6]. In this proving technique, system and its properties are specified in terms of mathematical logic. The verification of properties is done by discharging proof obligations generated by the system. If the proof is discharged for a statement, then it is known to be true and is said to be a theorem. The main advantage of theorem proving over model checking is that it can be used to verify the system having infinite states.

We have considered Event-B as a formal method for verification of our model. Event-B [7–9] is a formal technique, which is used to develop and formalize such system whose component can be modelled as discrete transition systems. It also provides refinement-based development of a complex model and has control systems within its scope. Event-B modelling can be used in various application areas like sequential programmes, concurrent programmes and distributed systems [10].

In this paper, we have developed formal model of distributed load migration mechanism using Event-B. Distributed system is a collection of autonomous systems connected by the network and they communicate with each other for the completion of common goal [11]. In this environment, the users submit the task at their sites for processing. The random arrival of tasks and their service order create the possibility that several sites may become heavily loaded and others may ideal or lightly loaded. It may degrade the performance of the whole system. Therefore, load distribution scheme is required for efficient use of resources and to enhance the performance [11–13]. In this paper, we have considered maximum load count value of site as the threshold value. This threshold value indicates maximum number of tasks that can be executed without affecting the performance of system. When a new task is submitted at site, the load count value of that site will be increased. When load count value of any site exceeds threshold value, then that site will become heavily loaded site. The load of this site should be transferred to idle or lightly loaded site. In order to find out low load site, heavily loaded site broadcasts load transfer request message to all sites. Site, whose load count value is lesser than threshold value, sends load reply message to sender. At any instance, lightly loaded site may receive number of load request messages from several heavily loaded sites. It will not send a reply message to all heavily loaded sites. It will send a reply only to that site first whose request for load transfer message arrived first. For ensuring ordered delivery of message, we have introduced a notion of causal order delivery [14]. The lightly loaded site will send a reply only to that site whose request for load transfer causally precedes the request from others. After receiving reply message from low load site, load from heavily loaded site will be transferred to it.

The remainder of this paper is organized as follows: Sect. 2 describes Event-B as formal method, Sect. 3 presents causal order broadcast, Sect. 4 outlines Event-B Model of causal order-based load distribution. This model consists of events like task submission event for submission of new task, enable and disable load transfer event for changing status of site when load count value increase and decrease from threshold value, broadcast and deliver event to formalize ordered delivery of load request message, reply message event, which models the sending of reply message to heavily loaded site and load reduction event to reduce load value. Finally, Sect. 5 concludes the paper.

## 2 Event-B

Event-B [14–18] is a formal technique, which captures complete system specifications on the basis of requirement and system behaviour. It can be expressed in form of states and events. It specifies the model in mathematical form. It defines mathematical structures as context and machine [19–21]. The context part of the model contains sets, constants and axioms, which are used to describe static properties of system. The dynamic part of the model is shown by machine part, which contains set of variables, invariants and events that modify the value of state variable when it triggers. The variables of model are constrained by invariants. The invariants of model which describe the properties of model should not be violated when an event occurs. The event contains guards which are necessary conditions for an event to trigger and list of actions. When all guards of an event become true, then set of actions written under it will be performed. The action of an event is expressed by substitution operation. It specifies how the state of system may change. The event may use local variables. The scope of that variable will be local to the event.

For ensuring correctness of the model, Event-B method requires to discharge all proof obligations generated by the model. Proof obligations serve to verify properties of the model.

There are several tools which support to write Event-B specifications. We have considered Rodin tool [22–24] for the development of our model. In order to discharge proof obligations generated by the model, there are various plug-ins which are provided by this tool. Event-B uses set-theoretic notations to specify the model. The syntax and detail description of Event-B notations can be found in [15].

## 3 Causal Order Broadcast

The formalization of causal relationship in distributed system was initiated by Lamport in [25]. Later, causal ordering of messages is proposed by Birman, Schiper and Stephenson [26]. The causal order property can be ensured by combining FIFO order and local order property [27].

FIFO order property says that if any site $Si$ broadcasts a message $Ma$ before broadcast another message $Mb$, then each receiving site delivers $Ma$ before $Mb$.

Local order property says that if any site $Si$ delivers message $Ma$ before broadcasting message $Mb$, then every receiving site delivers $Ma$ before $Mb$.

The causal order property says that if broadcasting of a message $Ma$ causally precedes broadcasting of a message $Mb$, then delivery of message $Ma$ at each site should be done before message $Mb$ (Fig. 1).

We can say that message $Ma$ causally precedes message $Mb$ if send event of message $Ma$; $send(Ma)$ at site $Si$ happened before message sending event $send(Mb)$ of message $Mb$ at site $Sj$. The causality of the message can also be related with receive event of the message. A message $Ma$ causally precedes $Mb$ if receive event of

**Fig. 1** Causal order
broadcast

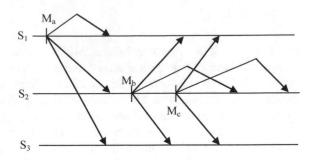

message *Ma* causally precedes the broadcast of *Mb*. As given in Fig. 1, broadcasting of message *Ma* causally precedes broadcasting of *Mb* and each recipient site delivers *Ma* before *Mb*. Similarly, broadcasting of message *Mb* causally precedes broadcasting of message *Mc* and each recipient site delivers *Mb* before *Mc*.

## 4 Event-B Model of Load Distribution Mechanism

We start with the distributed system model having a set of sites. Since there is no system-wide global clock or shared memory, the information from one site to other site is exchanged through messages. At any site, new task may be submitted. The task may be process or transaction which will perform some operation (reading or writing on data objects) at that site. Therefore, submission of new task will increase the load at that site. When the load count value exceeds a certain limit known as threshold value, then the performance of the system will degrade. In order to capture maximum throughput from the system, we need to distribute load from heavily loaded site to idle site in efficient manner.

In our model, the context part contains *SITE* and *MESSAGE* as carrier set. *TRANSFERSTATUS* and *TYPE* are declared as enumerated set. The set *TRANSFERSTATUS* represents load transfer status of site informs of disable and enable.

Initially, load transfer status of every site is disable because every site is underloaded. When load count value exceeds threshold value, then load transfer status will be set as enable. The set *TYPE* has element *LOAD_REQ* and *LOAD_REP* which is used to formalize type of message as load request and load reply, respectively. The machine part consists of variables, invariants and events. The description of variables is as follows (Fig. 2):

(i)   The variable *load_at_site* is specified as total function. It represents load count value of every site.
(ii)  The variable *loadtransferstatus* is a total function between site to *TRANSFERSTATUS*. Depending on the *loadcount* value, the load transfer status of every site may be either disable or enable.
(iii) The variable *threshold* value is declared as natural number.

> **Variables :**
>
> *load_at_site, loadtransferstatus, threshold_value, sender,*
> *cdeliver, messagetype, messagesent, delorder, corder,*
> *replymsgsent*

> **INVARIANTS**
>
> ***inv1:*** *load_at_site* $\in SITE \rightarrow \mathbb{N}$
>
> ***inv2:*** *loadtransferstatus* $\in SITE \rightarrow TRANSFERSTATUS$
>
> ***inv3:*** *threshold_value* $\in \mathbb{N}$
>
> ***inv4:*** *sender* $\in MESSAGE \nrightarrow SITE$
>
> ***inv5:*** *cdeliver* $\in SITE \leftrightarrow MESSAGE$
>
> ***inv6:*** *messagesent* $\subseteq MESSAGE$
>
> ***inv7:*** *messagetype* $\in messagesent \rightarrow TYPE$
>
> ***inv8:*** *corder* $\in MESSAGE \leftrightarrow MESSAGE$
>
> ***inv9:*** *delorder* $\in SITE \leftrightarrow (MESSAGE \leftrightarrow MESSAGE)$
>
> ***inv10:*** *replymsgsent* $\in (MESSAGE \leftrightarrow MESSAGE)$
>
> ***inv11:*** *ran(cdeliver)* $\subseteq dom(sender)$
>
> ***inv12:*** *dom(corder)* $\subseteq dom(sender)$
>
> ***inv13:*** *ran(corder)* $\subseteq dom(sender)$
>
> ***inv14:*** $\forall ss \cdot (ss \in SITE \wedge loadtransferstatus(ss)=enable \Rightarrow$
> *load_at_site(ss)>threshold_value)*

**Fig. 2** Variables and invariants of model

(iv) The variable *sender* is specified as partial function from *MESSAGE* to *SITE*. It models the sending of message *m* by site *s*.

(v) The variable *cdeliver* is declared as:

$$cdeliver \in SITE \leftrightarrow MESSAGE$$

The operator $\leftrightarrow$ defines the set of relations between *SITE* and *MESSAGE*. It represents causal delivery of messages at any site.

(vi) The variable *messagesent* represents the set of messages which have been sent.

(vii) The variable *messagetype* maps each sent message with its type. The type of message may be load request message or load reply message.

(viii) The variable *corder* models the causal relationship between messages. The mapping *(mmmm)* ∈ *corder* indicates that message *m* causally precedes message *mm*.

(ix) The delivery order of messages at any site is shown by *delorder*. It is declared as relation between site to set of ordered pair of messages. The mapping *ssm(mmmm)* ∈ *delorder* indicates that at site *ss* message, *m* is delivered before message *mm*.

(x) The variable *replymsgsent* models the sending of reply message corresponding to request message. The mapping *(m1mm2)* ∈ *replymsgsent* indicates that reply message *m1* has been sent corresponding to its request message *m2*.

## 4.1   Submission of Task

The event *TASK SUBMISSION* is given in Fig. 3. This event model the submission of task at any site *ss*. Every time when this event occurs, it increases the load count value of site by one. The action *act1* represents that load at site *ss* is incremented by one.

## 4.2   Enabling and Disabling Load Transfer Status

The event *ENABLE TRANSFER* updates load transfer status of site (Fig. 3). When load count value of any site exceeds threshold value, then this event updates load transfer status of that site as enable. The guard *grd2* ensures that load of site *ss* is greater than threshold value. The guard *grd3* ensures that load transfer status of site *ss* is disable. The action *act1* updates the load transfer status of site *ss* as enable.

The event *DISABLE TRANSFER* is given in Fig. 3. The guard *grd2* ensures that load count value of site *ss* is less than threshold value. The action *act1* set load transfer status of that site as disable.

## 4.3   Broadcasting and Delivery of Load Request Message

This event enables broadcasting of load request message to all sites (see Fig. 4). When the load count of any site exceeds its threshold value, then this site broadcast load request message to all sites. The purpose of the broadcasting request message is to know that which site is under loaded. The message that has not been sent is

```
TASK_SUBMISSION  ≙

ANY ss

WHERE

grd1:  ss∈ SITE

THEN

act1:  load_at_site(ss):=  load_at_site(ss)+1

END


ENABLE_TRANSFER  ≙

ANY ss

WHERE

grd1:  ss∈ SITE

grd2:  load_at_site(ss)>threshold_value

grd3:  loadtransferstatus(ss)=disable

THEN

act1:  loadtransferstatus(ss):=  enable

END
```

```
DISABLE_TRANSFER  ≙

ANY ss

WHERE

grd1:  ss∈SITE

grd2:  load_at_site(ss)<threshold_value

THEN

act1:  loadtransferstatus(ss):=  disable

END
```

**Fig. 3** Task submission, enable transfer and disable transfer event

ensured by guard *grd2* and *grd4*. The load transfer status of site *ss* is enable and is ensured through guard *grd3*. Due to the occurrence of this event, message *mm* will be broadcast (*act2*). The message *mm* will be added to *messagesent* set (*act1*). The action *act3* set the status of message *mm* as load request message. In this event, we are also ensuring ordered delivery of request message at sending site. The action *act4* ensures that all those messages which are sent by site *ss* will causally precede message *mm* (FIFO Order). The action *act5* specifies delivery of message *mm* at site *ss*. The action *act6* gives delivery order of message.

Delivery of load request message is given in Fig. 5. The guard *grd3* and *grd4* ensure that message *mm* has been sent but it is not delivered to site *ss*. The guard *grd5* is written as

$$\forall m \cdot (m \in MESSAGE \land (m \mapsto mm) \in corder \Rightarrow (ss \mapsto m) \in cdeliver)$$

It ensures that all messages *m* which causally precede message *mm* have already been delivered at site *ss*. Delivery of message *mm* at site *ss* is ensured by action *act1*. The action *act2* makes the delivery order of messages at site *ss*.

**BROADCAST** $\triangleq$

*ANY ss, mm*

**WHERE**

*grd1:  ss*$\in$ *SITE*

*grd2:  mm*$\notin$ *messagesent*

*grd3:  loadtransferstatus(ss)=enable*

*grd4:  mm*$\notin$ *dom (sender)*

**THEN**

*act1:   messagesent*$:=$ *messagesent*$\cup${mm}

*act2:   sender*$:=$*sender*$\cup$ {mm$\mapsto$ss}

*act3:   messagetype(mm)*$:=$*LOAD_REQ*

*act4:*   corder$:=$corder $\cup$((sender$\sim$[{ss}] $\times$ {mm})

         $\cup$(deliver[{ss}] $\times$ {mm}))

*act5:   cdeliver*$:=$ *cdeliver*$\cup$ {ss$\mapsto$ mm}

*act6:   delorder* $:=$ *delorder* $\cup${ss$\mapsto$(cdeliver[{ss}] $\times$ {mm})}

**END**

**Fig. 4** Broadcast event

## 4.4  Sending of Reply Message

The event *REPLY* is given in Fig. 6. This event models sending of load reply message (*LOAD_REP*) by those sites whose load count value is lesser than threshold value. The load request message *mm* has been received by site *s* is ensured by guards *grd3 and grd4*. Site may receive number of load request messages from several heavily loaded sites but it will send a reply only to that site whose load request message *LOAD_REQ* message causally precedes other requests. The guard *grd5* ensures that load request message *mm* causally precedes all load request message *msg*. Therefore, site *s* will send load reply message *m* corresponding to load request message *mm* only. The load count value of site *s* is lesser than threshold value and is ensured by guard *grd6*. The reply message *m* has not been previously sent is ensured through guard *grd7 and grd8*. The action *act1* ensures sending of message *m* by site *s*. The action *act2* adds the message *m* to *messagesent* set. The action *act3* set the status

---

**DELIVERY** ≙

*ANY ss,mm*

*WHERE*

*grd1:* $ss \in SITE$

*grd2:* $mm \in MESSAGE$

*grd3:* $mm \in dom(sender)$

*grd4:* $(ss \mapsto mm) \notin cdeliver$

*grd5:* $\forall m \cdot (m \in MESSAGE \wedge (m \mapsto mm) \in corder \Rightarrow$
$(ss \mapsto m) \in cdeliver)$

*THEN*

*act1:* $cdeliver := cdeliver \cup \{ss \mapsto mm\}$

*act2:* $delorder := delorder \cup \{ss \mapsto (deliver[\{ss\}] \times \{mm\})\}$

*END*

---

**Fig. 5** Deliver event

of message *m* as *load reply (LOAD_REP)*. The action *act4* makes the entry of reply message *m* corresponding to request message *mm*.

## 4.5 Load Reduction Event

This event model the load reduction from heavily loaded site (Fig. 7). The guard *grd2* ensures that site *ss* is heavily loaded because its load count value is greater than threshold value. Guards *grd3, grd4 and grd5* ensure that site *ss* has received the load reply message *m*. Receiving of reply message also indicates that there is some site whose load count value is less than threshold value. Request message *mm* sent by site *ss* is ensured through guards *grd6, grd7 and grd8*. The guard *grd9* ensures that message *m* is reply message of request message *mm*. Due to the occurrence of this event, load count value of site *ss* is reduced by one (*act1*). After the reduction of load, it will be submitted at low load site in the form of task through TASK SUBMISSION event.

> **REPLY** ≜
>
> *ANY s, mm, m*
>
> **WHERE**
>
> *grd1:*  $s \in SITE$
>
> *grd2:*  $mm \in messagesent$
>
> *grd3:*  $messagetype(mm)=LOAD\_REQ$
>
> *grd4:*  $s \mapsto mm \in cdeliver$
>
> *grd5:*  $\forall msg\cdot(msg \in messagesent \land messagetype(msg)$
> $=LOAD\_REQ \Rightarrow (mm \mapsto msg) \in corder)$
>
> *grd6:*  $load\_at\_site(s)<threshold\_value$
>
> *grd7:*  $m \in MESSAGE$
>
> *grd8:*  $m \notin dom(sender)$
>
> **THEN**
>
> *act1:*  $sender:=sender \cup \{m \mapsto s\}$
>
> *act2:*  $messagesent:=messagesent \cup \{m\}$
>
> *act3:*  $messagetype(m):=LOAD\_REP$
>
> *act4:*  $replymsgsent:=replymsgsent \cup \{m \mapsto mm\}$
>
> **END**

**Fig. 6**  Reply event

## 5  Conclusion

Distributed systems provide tremendous processing capacity. In order to maximize the performance of the system, good load transfer or task migration schemes are required. The random arrival order of task and its random system service may create the situation that all resources and systems may not properly be utilized. Due to uneven load distribution, few of the sites may become heavily loaded and others may be ideal. We have introduced causal order delivery of load transfer request message which ensures ordered service of load request message.

In this paper, a formal development of causal order-based load distribution mechanism is done. Formal methods are mathematical techniques to verify the correctness of system properties. We have considered Event-B as a formal method for the devel-

**Fig. 7** Load reduction event

```
LOAD_REDUCTION  ≙

ANY ss, m, mm

WHERE

grd1:  ss∈ SITE

grd2:  load_at_site(ss)>threshold_value

grd3:  m∈ messagesent

grd4:  messagetype(m)=LOAD_REP

grd5:  ss↦m∈ cdeliver

grd6:  mm∈ messagesent

grd7:  mm↦ss∈ sender

grd8:  messagetype(mm)=LOAD_REQ

grd9:  (m↦mm)∈ replymsgsent

THEN

act1:  load_at_site(ss):=load_at_site(ss)−1

END
```

opment of our model. In Event-B model, the properties of model are specified through invariants. These invariants should not be violated during the execution of the model. Event-B model generates proof obligations, and we need to discharge all proofs generated by it. A total of 42 proof obligations are generated by the model out of which 30 proofs are discharged automatically whilst 12 proofs are discharged interactively. In order to ensure correctness, we have added the following invariants:

$$\text{ran(cdeliver) (dom(sender)} \dots \text{(inv11)}$$

$$\text{dom(corder) (dom(sender)} \dots \text{(inv12)}$$

$$\text{ran(corder) (dom(sender)} \dots \text{(inv13)}$$

$$! ss(ss \in SITE \& loadtransferstatus(ss) = enable)G \ load \ at \ site(ss) > (threshold \ value) \dots (inv1$$

The invariant inv11 ensures that messages which are delivered should be a subset of messages which have been sent. Similarly, invariants inv12 and inv13 ensure that messages for which ordering is maintained (causal order) should be a subset of sent messages. The invariant inv14 ensures that if load transfer status of any site is enable, then load of that site exceeds threshold value of that site.

The invariants and proofs of model give a clear insight of model. In future, we plan to strengthen invariants and add fault-tolerance property to this model.

# References

1. Bjrner, D.: Logics of formal specification languages. Comput. Inform. **22**(1–2), This double issue contains the following papers on B, CafeOBJ, CASL, RAISE, TLA+ and Z (2003)
2. Bjrner, D.: Special double issue on formal methods of program development. Int. J. Softw. Inform. **3** (2009)
3. Shankar, N.: Combining theorem proving and model checking through symbolic analysis. In: Proceeding of CONCUR '00, vol. 1877, pp. 1–16. LNCS, Springer (2000)
4. Fitzgerald, J., Larsen, P.G.: Modelling Systems—Practical Tools and Techniques in Software Development. Cambridge University Press, Cambridge, UK, Second edition (2009)
5. Clarke, E., Zhao, X.: A theorem prover for mathematica. In automated deduction-CADE-II. In: 11th International Conference on Automated Deduction, pp. 761–763. Saratoga Springs, New York, 15–18 June 1992
6. Clarke, E., Zhao, X.: A theorem prover for Mathematica. Math. J. (1993)
7. Abrial, J., Butler,M., Hallerstede,S., Voisin, L.: An open extensible tool environment for Event-B. In: Liu, Z., He, J. (eds.) ICFEM, Lecture Notes in Computer Science, vol. 4260, pp. 588–605. Springer (2006)
8. Abrial, J.R.: Modeling in Event-B: System and Software Engineering. CambridgeUniversity Press (2010)
9. Abrial, J.R., Hallerstede, S.: Refinement, decomposition, and instantiation of discrete models. Appl. Event B Fundam. Inform. **77**(1–2), 1–28 (2007)
10. Butler, M.: An approach to the design of distributed systems with B AMN. In: Bowen, J.P., Hinchey, M.G., Till, D. (eds.) ZUM, Lecture Notes in Computer Science, vol. 1212, pp. 223–241. Springer (1997)
11. Singhal, M., Shivratri, N.G.: Advanced Concepts in Operating Systems. Tata McGraw-Hill Book Company (2012)
12. Lazowska, D.E., Zahorjan, J.: Adaptive load sharing in homogeneous distributed systems. IEEE Trans. Softw. Eng. **12**(5), 662–675 (1986)
13. Lazowska, D.E., Zahorjan, J.: A Comparison of receiver-initiated and sender-initiated adaptive load sharing. Perform. Eval. **6**(1) 53–68 (1986)
14. Yadav, D., Butler, M.: Application of Event B to global causal ordering for fault tolerant transactions. In: Proceeding of Workshop on Rigorous Engineering of Fault Tolerant System, REFT05, Newcastle upon Tyne, pp. 93–103, 19 July 2005
15. Yadav, D., Butler, M.: Rigorous design of fault-tolerant transactions for replicated database systems using Event B. In: Butler, M., Jones, C.B., Romanovsky, A, Troubitsyna, E. (eds.) Rigorous Development of Complex Fault-Tolerant Systems. Lecture Notes in Computer Science, vol. 4157, pp. 343–363. Springer, Heidelberg (2006)
16. Yeganefard, S., Butler, M., Rezazadeh, A.: Evaluation of a guideline by formal modelling of cruise control system in Event-B. Proc. NFM **2010**, 182–191 (2010)
17. Liu, J., Liu, J.: A formal framework for hybrid Event B. Electron. Notes Theor. Sci. **309**(2014), 3–12 (2014) (Elsevier)
18. Suryavanshi, R., Yadav, D.: Formal development of byzantine immune total order broadcast system using Event-B. In: Andres, F., Kannan, R. (eds.) ICDEM 2010. LNCS, vol. 6411, pp. 317–324. Springer, Germany (2010)
19. Hallerstede, S., Leuschel, M.: Experiments in program verification using Event-B. Form. Asp. Comput. **24**, 97–125 (2012)

20. Suryavanshi, R., Yadav, D.: Rigorous design of lazy replication system using Event-B. In: Communications in Computer and Information Science, vol. 0306, pp. 400–411. Springer, Germany (2012). ISSN 1865-0929

21. Suryavanshi, R., Yadav, D.: Modeling of multiversion concurrency control system using Event-B. In: Federated Conference on Computer Science and Information systems (FedCSIS), Poland, indexed and published by IEEE, pp. 1397–1401, 9–12 Sept 2012. ISBN 978-83-60810-51-4

22. Banach, R.: Retrenchment for Event-B: usecase-wise development and Rodin integration. Form. Asp. Comput. **23**, 113–131 (2011)

23. Abrial, J.R., Cansell, D., Mery, D.: A mechanically proved and incremental development of ieee 1394 tree identify protocol. Form. Asp. Comput. **14**(3), 215–227 (2003)

24. Metayer, C., Abrial,J.R., Voison, L.: Event-B language. RODIN deliverables 3.2. http://rodin.cs.ncl.ac.uk/deliverables/D7.pdf (2005)

25. Lamport, L.: Time, clocks, and the ordering of events in a distributed system. Commun. ACM **25**(7), 558–565 (1978)

26. Birman, K., Schiper, A., Stephenson, P.: Lightweight causal and atomic group multicast. ACM Trans. Comput. Syst. **9**(3), 272–314 (1991)

27. Yadav, D., Butler, M.: Formal specifications and verification of message ordering properties in a broadcast system using Event B. In: Technical Report, School of Electronics and Computer Science, University of Southampton, Southampton, UK (2007)

# An IOT-Based Architecture for Crime Management in Nigeria

Falade Adesola, Sanjay Misra, Nicholas Omoregbe, Robertas Damasevicius
and Rytis Maskeliunas

## 1 Introduction

A crime is an act of wrongdoing that merit community condemnation and punishment usually by way of payment of fine or imprisonment [1]. There are numerous examples of crime, but this study is confined to rape, armed robbery, murder, kidnapping, and ritual killings, which are more prevalent in Nigeria [2]. Rate of crime is inversely proportional to the rate of economic development of any nation [3, 4]. This implies that no nation can experience any meaningful development with a high rate of crime. Hence, it is imperative for the government of any nation to curtail or eradicate this monster from society.

The activities and the dastard effect of this menace on the socioeconomic development of any nation are worrisome [5]. With Nigeria being the case study, the criminal activities have been on the increase in recent times as a result of economic downturn and recession in the country. This has had negative effects on the socioeconomic development of the country. Billions of dollars have been lost by Nigeria as a result of this ugly trend [6]. The report on Nigeria watch [7] showed the most crime fatalities tending to occur in major cities of the country such as Lagos and Rivers.

F. Adesola · S. Misra (✉) · N. Omoregbe
Center of ICT/ICE Research, CUCRID Building, Covenant University, Ota, Nigeria
e-mail: sanjay.misra@covenantuniversity.edu.ng

F. Adesola
e-mail: adesola.falade@covenantuniversity.edu.ng

N. Omoregbe
e-mail: nicholas.omoregbe@covenantuniversity.edu.ng

R. Damasevicius · R. Maskeliunas
Kaunas University of Technology, Kaunas, Lithuania
e-mail: robertas.damasevicius@ktu.lt

R. Maskeliunas
e-mail: rytis.maskeliunas@ktu.lt

© Springer Nature Singapore Pte Ltd. 2019
R. K. Shukla et al. (eds.), *Data, Engineering and Applications*,
https://doi.org/10.1007/978-981-13-6351-1_19

In addition, a graph showing the various incidence of armed robbery dwindled over the years, and the report concluded that the fatalities have remained high since then. To this end, our research question is "What are the steps to be taken to eradicate or reduce to barest minimum the growing rate of crime in Nigeria?" Application of Internet of Things and Big Data ICT technologies have shown great potentials from recent studies [8–12] and so we believe it can give us an optimal solution to this problem where criminals and criminal activities can be monitored, tracked, and detected real-time online [4]. This will go a long way to assist security agents to do their job efficiently and productively. Therefore, the aim of this work is to harness the current state of technology in the Internet of Things and Big Data technologies and develop a model that can be used to track and monitor crime real-time online.

## 2 Related Works

According to IEEE IoT community, Internet of Things (IoT) can be described as a collective network of sensors and smart objects that are self-configuring and adaptive for the purpose of identification, communication, sensing, and data collection as well as interaction with humans [13]. The use of IoT in crime detection and monitoring will bring comfort to security agents because of its various applications such as real-time monitoring, crime prevention, crime information management system, and implementation of smart cities [14].

Manual-based approach to crime recording and documentation has been the norm and practices by the local police in Nigeria. When a crime is reported or brought to the police station, the usual practice is that an incident sheet is given and the person involved is asked to write a statement in the incident sheet. After this, the police officer in charge is required to keep the recorded statement in the incident case file [15].

In 2011, a Crime Tracker was developed and deployed in Enugu state in Nigeria, where it was test run and adopted by the police command in that state for managing criminal records [15]. It has really helped the police authority in that command to be more efficient in managing criminals and criminalities in the state. The software is popular among the police in Enugu state, Nigeria. However, Crime Tracker has the following shortcomings: First, the backend was built on a relational database model, this is a limitation if we have to extend the functionalities to Big Data. This is because crime data is huge and in different forms (videos, audio, text, emails, images, etc.). Second, Crime Tracker is not web-based; it is not accessible by tablets and mobile devices.

The study in [16] proposed the use of surveillance CCTV cameras and smart-phones equipped with GPS technology to track and monitor mobile phone user. The surveillance mobile app installed on the smartphone can be launched at will to start video streaming and sending this to the centralized monitoring server, which in turn sends to the police authority for action. The limitation of their work is that it does not work on all mobile phone platforms.

The study in [17] presents a ubiquitous crime prevention system that leverages on Internet access and a mobile app. The mobile app provides information based on the factors that could cause crime occurrence within an area. The mobile app is connected to a big data analytic engine that could analyze public big data such as credit card usage, pedestrian flows, smartphone usage, etc. The resulting big data is then analyzed using spatial statistical analysis to produce factors affecting crime occurrence and then provide crime information to the general public through the Internet or a mobile app.

The study in [18] developed a model for smart crime detection using IoT that is able to detect crimes in real time by analyzing the human emotions. It serves as a tool for both police agencies to determine crime as well as for citizens to be on the safe side of the places they live in. The limitation, however, is that the use of only wearable sensing devices was proposed and this could be subjected to removal during striping when attacked by kidnappers.

As for the study in [14], a smart community using wireless communication and ubiquitous sensing technologies to connect smart homes in a local community is proposed. It provides useful functions for the local residence—neighborhood watch and pervasive healthcare with a limitation of no capability to detect and track criminals.

The study in [19] proposed a system that identifies deception in communication through emails about criminal activities. In the study, decision tree classification data mining technique was applied to detect deception in suspicious emails. Deception is usually characterized by reduced frequency of first-person pronouns and exclusive words, elevated words as well as the use of action verbs. They applied this model of deception to a large email dataset and then applied decision tree classification techniques. The decision tree that was generated was then used to classify the email as fraudulent or not. However, the gap cited in their work is that it does not take care of real-time monitoring of criminals and only using emails to detect crimes is not enough.

As a result of a variety of gaps cited in [15–18], we are motivated to fill these gaps using current technologies in crime recording and tracking in Nigeria whose trends have been on the increase recently. By using big data technologies and IoT, it becomes very easy to derive and mine security intelligence from the vast volume of crime data [13].

## 3 Conceptual View of the Proposed Model

In this section, the conceptual model of the architecture is depicted in Fig. 1.

From Fig. 1, the following components can be observed: (i) the emotion state-sensing module that has the ability to sense the emotions of the user through a body-centric RFID enabled smart sensor, which is worn by the user (ii) there is also the emotion state recording module that records and stores the type of crime detected (iii) there is a crime detection module, which is responsible for sending alerts to the stakeholder concerned, for instance, the police authority or relative of the user (iv)

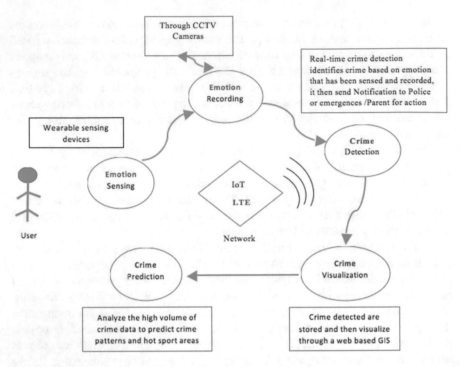

**Fig. 1** The proposed crime prediction and monitoring model

furthermore, there is the crime visualization module, which stores the detected crime in a repository and then visualizes it through a web-based Geographical Information System (GIS) (v) finally, the fifth component is the crime prediction module for making crime predictions through the analysis of high volume of crime data. From Fig. 1, every person being tracked is equipped with a wearable or implantable device. The sensor here is a body-centric RFID enabled smart sensor that is capable of sensing the emotions of the user. This will be interconnected with and monitored through programmable CCTV cameras that have the capability of recording 36 emotional states of humans. If the person is attacked or in a dangerous situation, his/her emotion changes and this is sensed and recorded through the CCTV cameras and the emotion recording component module to get the actual crime being committed. In order to improve the accuracy of the type of crimes detected, the user emotions are recorded through emotions programmable CCTV cameras [18]. The crime detection component identifies crimes that are based on the emotions that were sensed and recorded. Alerts and notifications are then sent to the Police authority patrol vehicle for urgent action to be taken after the crime is reported. The crime visualization component stores the detected crimes in a database and then visualizes it through a web-based Geographical Information System (GIS). Crime prediction works by analyzing a big volume of crime data, prediction can then be made for future crimes and area of crime hot spots can be avoided, and also help police to focus on the most problematic regions.

## 4 Implementation

In this section, the proposed model is implemented as a proof of concept. The system operates as follows.

A startup and welcome screen are first displayed as depicted in Figs. 2 and 3. In Fig. 3, the user is expected to enter a valid user name and password in order to proceed into the system. On successful login, the Administrator/user is provided with a menu option to choose from as depicted in Fig. 4. The options include criminals in a locality; crime Type; criminals record search, and criminal's search by description. If a user selects criminals in a locality, he would also have to specify the locality so as to retrieve relevant records. The user can then go on to select a crime type (e.g., rape, arm robbery, kidnapping, etc.) and click OK as depicted in Fig. 5. If there are records based on the search, the record is displayed as given in Fig. 6.

It is important to note that in the implementation of the web-based GIS in the above model, Microsoft Windows 8 was the operating system, on which the machine was running. MySQL and NoSQL were used as backend database for storing criminal records and PHP, HTML, NodeJS platform was used for the implementation of the proposed model. These development tools were chosen because of their simplicity

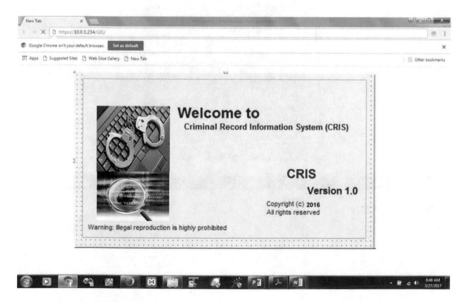

**Fig. 2** Welcome screen

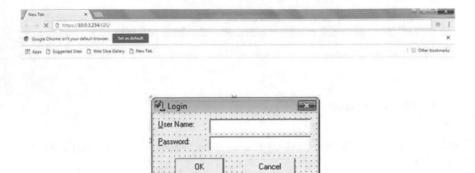

**Fig. 3** Login screen

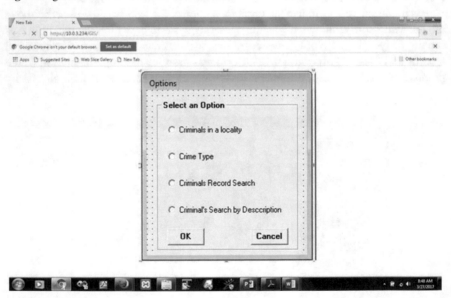

**Fig. 4** Options Dialog

**Fig. 5** Criminal Mode of operation

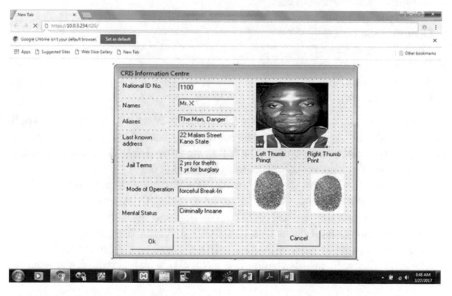

**Fig. 6** Information form about criminal

**Table 1** Comparison of our proposed application with an existing system

| Functionalities | Crime tracker | Proposed application (CRIS) |
|---|---|---|
| GPS required | No | Yes |
| Costing | Expensive | Cheap |
| Ease of tracking criminals | Yes | Yes |
| Web-based GIS | No | Yes |
| Interoperability | No | Yes |
| Support NoSQL | No | Yes |

and flexibility in coding, easy integration, and deployment. This will serve as a tool for the police authority and go a long way in assisting security agents in keeping track and saving criminal records in the database, and also is able to speedily retrieve necessary information about criminals and any other offenders when needed.

The designed system was implemented, tested, evaluated, and interpreted as discussed in this section. The system captured the following information on criminals:

- The names and the aliases used
- Crime type carried out
- Operation mode
- The Jailed term issued
- Committed last crime date
- State of mental status
- Criminal sex, etc.

The welcome screen, login page, and the rest of the interactive screens in the application provide the step-by-step process for tracking, record keeping, and searching and criminal records management. The following table (Table 1) shows the comparative analysis of our proposed system with an existing one.

From Table 1, it can be observed that the comparison was done based on functionality (i.e., GPS required, Costing, Ease of Tracking Criminals, web-based GIS, Interoperability, and Support for NoSQL). As depicted in Table 1, our application requires GPS functionality; it is web-based GIS (implying a wider audience coverage), and also is able to work with other softwares such as fingerprint reader (interoperability). Our software also supports the NoSQL, which is requisite to handling big data records.

## 5 Result and Discussion

The results obtained showed that the model and the application developed can be used to track and manage crime in Nigeria and the benefits of the proposed system that actually makes it different from the existing one are summarized as follows:

1. The application we are proposing here is comparatively cheaper to develop and use comparing with the existing systems, this is because it was developed using an open source platform, which is free to use. The existing one was developed using a proprietary platform, visual studio.net that is a paid license platform of development. Also in terms of simplicity and flexibility in usage, this is a system to be reckoned with [15].
2. The proposed application is simply convenient in easily tracking and locating criminals. This was confirmed during the testing stage where it was deployed [9].
3. This automated system will help in facilitating the record keeping of criminals for future references and if finally adopted, will make it difficult for criminals to escape authority. This is because the fingerprint and other biometric data of criminals are already stored in the system as shown in Fig. 6.

# 6 Conclusion and Future Work

This study has afforded us the opportunity to make contributions toward a crime free society by using IoT and Big Data technologies in curbing the growing crime rates in the country. The proposed crime detection and monitoring model and the software application evolved, if adopted by the government will go a long way in assisting our security agents to be more efficient and proactive in detecting crimes and hot spots areas in the country. Future work should, however, be focused on the maintainability of the software developed since this is a work in process and often comes with heavy cost. Software maintainability usually prolongs the lifespan of any software. However, future research work tends to show a detailed maintainability and security model. Stringent security measures should be put in place to prevent third party and cybercriminals from hacking the system.

**Acknowledgements** We acknowledge the support and sponsorship provided by Covenant University through the Centre for Research, Innovation and Discovery (CUCRID).

# References

1. Ellis, A.: Theories of punishment. In: Encyclopedia of Criminology and Criminal Justice, pp. 5184–5192. Springer, New York (2014): Morris, S.V.: Crime and Punishment in Africa. The Encyclopedia of Crime and Punishment (2016)
2. Renold, A.P., Rani, R.J.: An internet based RFID library management system. In: IEEE Conference on Information & Communication Technologies (ICT), pp. 932–936 (2013)
3. Osabiya, B.J.: Ethnic militancy and internal terrorism on Nigeria's national security. Int. J. Dev. Confl. **5**, 59–75 (2015)
4. Li, X., Chen, J., Lin, X.: Smart community: an internet of things application. Commun. Mag. **49**, 68–75 (2011)

 5. Abioro, T., Adefeso, H.A.: The menace of poverty and the challenges of public policy making in Nigeria. J. Sustain. Dev. **9**, 177 (2016)
 6. Dlodlo, N., Mbecke, P., Mofolo, M., Mhlanga, M.: The internet of things in community safety and crime prevention for South Africa. In: Innovations and Advances in Computing, Informatics, Systems Sciences, Networking and Engineering, pp. 531–537. Springer International Publishing (2015)
 7. NSRP: Nigeria Watch Project: Fifth Report on Violence (2015). http://www.nigeriawatch.org/media/html/NGA-Watch-Report15Final.pdf. Accessed 15 Mar 2017
 8. Xiao, L., Wang, Z.X.: Internet of things: a new application for intelligent traffic monitoring. Syst. **6**, 887–894 (2011)
 9. Lopes, N.V., Santos, H., Azevedo, A.I.: Detection of dangerous situations using a smart internet of things system. In: New Contributions in Information Systems and Technologies. Springer International Publishing, pp. 387–396 (2015)
10. Jalali, R., El-Khatib, K., McGregor, C.: Smart city architecture for community level services through the internet of things. In: 18th International Conference on Intelligence in Next Generation Networks (ICIN), pp. 108–113 (2015)
11. Suh, D.H., Song, J.H.: Establishing crime prevention systems based on internet of things and associated spatial urban factors (2016)
12. Kim, G.H., Trimi, S., Chung, J.H.: Big-data applications in the government sector. Commun. ACM **57**, 78–85 (2014)
13. Botta, A., De Donato, W., Persico, V., Pescapé, A.: Integration of cloud computing and internet of things: a survey. Futur. Gener. Comput. Syst. **56**, 684–700 (2016)
14. Swapnali, R., Rohini, B., Kaustubh, B., Mahendra, S.: Crime monitoring and controlling system by mobile device. Int. J. Recent. Innov. Trends Comput. Commun. **3**, 123–126 (2014)
15. Tae-Heon, M., Sun-Young, H., Sang-Ho, L.: Ubiquitous crime prevention system (UCPS) for a safer city. In: 12th International Conference on Design and Decision Support Systems in Architecture and Urban Planning (2014)
16. Byun, J.Y., Nasridinov, A., Park, Y.H.: Internet of things for smart crime detection. Contemp. Eng. Sci. **7**, 749–754 (2014)
17. Pulim, P.R., Rajesh, D.: Data analytics applied for crime identification over IOT. Int. J. Sci. Res. **5** (2016)
18. Cai, H., Xu, L.D., Xu, B., Xie, C., Qin, S., Jiang, L.: IoT-based configurable information service platform for product lifecycle management. IEEE Trans. Industr. Inf. **10**, 1558–1567 (2014)
19. Jara, J., Zamora-Izquierdo, M.A., Skarmeta, A.F.: Interconnection framework for mHealth and remote monitoring based on the internet of things. IEEE J. Sel. Areas Commun./Suppl. **31**, 47–65 (2013)

# A Comparative Analysis of Techniques for Executing Branched Instructions

Sanjay Misra, Abraham Ayegba Alfa, Kehinde Douglas Ajagbe, Modupe Odusami and Olusola Abayomi-Alli

## 1 Introduction

Though in well-defined pipeline stage that is at an appropriate instant, pipelining allows enormous number of instructions to be performed simultaneously [1]. For better performance to be achieved, pipelining is used to overlay the execution of instructions. Instruction-level parallelism (ILP) describes the prospective overlay for different sets of instructions [1]; simply because these instructions can be carried out in parallel [2]. As a result of complication in nature and size of instructions constructs, the circumstances curbing utilizing of ILP in compilers and processing elements keep on increasing. Some of the drawbacks observed include greater part supports only various problems of straight instructions, also prediction analysis is needed to convert branch/conditional instructions to straight instructions, and this usually result in stall of memory/compiler system. One inlet and outlet access in loop body (which is not flexible to allow new instructions scheduling interrupts until the inlet or outlet is detected) in basic block (BB) architecture is available [3]. Among the instructions in basic block architecture, there is primarily tiny or no overlay. By using different architectures and techniques, various forms of parallelism can be exploited.

S. Misra (✉) · M. Odusami · O. Abayomi-Alli
Center of ICT/ICE Research, CUCRID Building, Covenant University, Ota, Nigeria
e-mail: sanjay.misra@covenantuniversity.edu.ng

M. Odusami
e-mail: modupe.odusami@covenantuniversity.edu.ng

O. Abayomi-Alli
e-mail: olusola.abayomi-alli@covenantuniversity.edu.ng

A. A. Alfa · K. D. Ajagbe
Kogi State College of Education, Ankpa, Nigeria
e-mail: abrahamsalfa@gmail.com

K. D. Ajagbe
e-mail: dougajagbe@gmail.com

© Springer Nature Singapore Pte Ltd. 2019
R. K. Shukla et al. (eds.), *Data, Engineering and Applications*,
https://doi.org/10.1007/978-981-13-6351-1_20

Techniques such as single instruction, multiple data (SIMD), very long instruction Word (VLIW), and the superscalar execution could be used to attain parallel execution at the instruction level. Practical usage of parallelism intrinsic in algorithms with high level can be greatly achieved through multicore architectures [4]. This paper shows the differences between the predication process and the two-way loop process for branched instructions. The rest of the paper is structured as follows: the review of branched instruction execution and execution cycle instructions are presented in Sect. 2. Design methodology and various technique used are discussed in Sect. 3. Results and discussion are presented in Sect. 4. Conclusion of the paper is done in Sect. 5.

## 2   Review of Branched Instruction Execution

Unrolling loop technique is one of the most utilized techniques for executing branched instruction. In order to exploit parallelism by bypassing branch instructions, many enlarged basic blocks are contained in the unrolling of loop techniques. Enlarging basic block can be referred to as monotonous activities like loops using an algorithm to attain an efficient and small-scale code. Provided the bound variables are defined at compile time and unchanged, however, a loop body can be repeated for a couple of times using this algorithm [4, 5].

Figure 1 depicts the beginning of every iteration for unrolling loop, and no control exists for instruction. Although, there exist more compact in the loop form. In view of stable bound loops making the control instruction by the loop form naturally irrelevant, total iterations are determined at compile time. Unrolling loop appropriated the issue of instruction control. The unrolling loop has the capability to duplicate the body of loop $n$ time with iteration step $n$ incrementation of induction variable by a factor of two for every iteration and unrolling by a factor of two is shown in Fig. 1. For parallel execution, the loop body yields $n$ times longer and higher number of instructions [4–6].

With the view of increasing the capacity for parallelism, unrolling loop expands basic blocks. The loop body is duplicated n times and the counter of the loop changes

**Fig. 1** Unrolling loop
architecture and code [6]

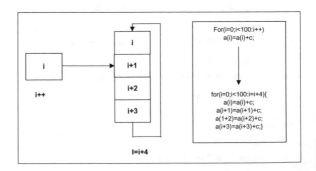

by the step n with the unrolling loop. One major drawback of the unrolling loop technique is the difficulty in increasing the code size which does not apply to the outer loop. Considering compiler-time scheduling, the advantage of benefits from parallelism cannot be taken by unrolled adjacent. Out-of-order execution advantage taken by dynamic scheduling make the successive iteration run in parallel alongside the current one.

## 2.1 Instructions Execution Cycle

Instruction execution cycle is described as the implementation of a single basic block of instructions on compiler sectioned to different operations. Successive instruction address is being maintained by the program counter. Specific instruction to be carried out is preserved by the instruction queue. Fetch, decode, and execute are the three principal steps for carrying out a compiler instruction. If a memory operand is used by the instruction, two additional steps are required, and these are: fetch operand and store output operand [7]. Descriptions of each of the steps are as follows:

Fetch: Instruction is being fetched from the instruction queue by the control unit and there is an increment in instruction pointer (IP).

Decode: To confirm what the instruction does, the function of instruction is decoded the control unit. Signals are forwarded to arithmetic logic unit (ALU) defining the action to be executed as input operands of instruction are being accelerated.

Fetch operands: The control unit indulges the read operation to recollect operand and duplicate it to internal registers as long as the instruction desires an input operand saved on memory. User programs are hidden from internal registers.

Execute: Internal registers and named registers are used as operands as the ALU executes instruction and forward the outcome to memory and/or named registers. ALU updated the status flags by reading the state of processor information.

Store output operand: The advantage of a write operation is used by control unit to save data as long as the location of the output operand is in the memory.

Loop unrolling hardware caching: Hardware made up of components such as a negative branch displacement, stack-based approach, and a comparator (use to detect) are used to classify loop bodies. Information for loop linked is recorded on the loop stack upon loop entry discovery during commit time [8]. The total of successful iterations per visit and for every iteration in-loop branch log is accounted for by this dynamic information [8, 9].

## 3 Methodology

The design is divided into three stages and these include First stage: Choose the pipelining and two-way loop as the materials for study. Second stage: Conduct branched or conditional construct for each of the techniques is selected in the first phase.

**Table 1** Setup of two-way loop technique [1]

| Component | Characteristics |
| --- | --- |
| ILP | Overlapping/interleave of instructions that support multiple issues |
| Model | Object/rapid iterative |
| Program style | Code motion/transformation. Two-way algorithm |
| GUI | Microsoft Visual Basic programming language |
| Unrolling loop | Replicating loop bodies into many other independent sub-loops |

Third stage: Evaluation of the performance of each technique using utilization, execution rate, and regression coefficients indices.

## 3.1 A Two-Way Loop Technique Setup

The TWL technique offers a breakdown of the process of making more ILP accessible to basic block instructions in loop structure as shown in Table 1.

## 3.2 An Algorithm for Two-Way Loop

Various issues/simultaneous instructions executions of branch and straight path of loops is supported by two-way loop algorithm [10]. Three main things are achieved by the TWL algorithm and these include transforms of control dependences into data dependences, increasing in parallelism exploited by layers of instructions in a compiler, and transforms unrolling of loop technique by increasing basic block, respectively, in order to increase the chance of exploiting parallelism, thereby granting several branched instructions to be executed. For a competent and more solid form of the resulting code to be achieved, repeated operations of an algorithm are regularly revealed as a loop. The steps are [10]:

1. Establish conditional branch instructions //over a number of loop unrolling.
2. Step A instruction is converted to predicate defining instructions //instructions that fix a certain degree known as a predicate.
3. Modification of instructions associated to both branch and straight constructs into predicate instructions //according to the degree of the predicate, both of them carried out instruction.
4. Fetch and execute predicated instructions regardless of the degree of their predicate//over a number of loops unrolling.
5. Instructions retirement stage.
6. For predicate degree = TRUE //progress to the next and last pipeline phase.

**Fig. 2** Framework of two-way loop mechanism [10]

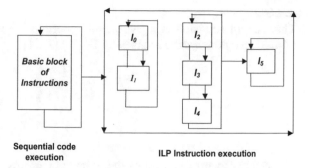

Sequential code execution

ILP Instruction execution

7. For predicate degree = FALSE //nullified: there is no need to write back the results that is produced and hence lost.

Competent supervision of instruction control, conditional instructions, and storage for predicate values of different basic blocks loops are supported by the algorithm [10].

### 3.3 Framework of Branched Instructions Mechanism

The embedded algorithm in the new design enabled the processing unit of the mechanism, which has fixed characteristic of ILP architecture to control multiple instruction sets, which could be conditional or branch instructions and process them at the shortest CPU cycles time [10]. The result is the capability to handle instructions such as branch and straight instructions, parallelism exploitation among instructions, and the speed of instructions execution as shown in Fig. 2.

### 3.4 Metrics of Performance

The comparisons of the two branched instruction processing techniques are measured such as the closeness of execution results. ILP impact is measured using the speedup in execution time and is expressed by Eq. (1)

$$ILP\ Speedup = \frac{T0}{T1}.\qquad(1)$$

where

T0  previous technique execution time
T1  TWL technique execution time.

*Amdahl's Law*

The states of Amdahl's law gave the total speedup of a particular component or rate of used up by the system as given by Eq. (2) [7, 11]

$$S = \frac{1}{(1-f) + \frac{f}{k}}. \tag{2}$$

where

S   speedup for total system.
F   section of work executed by the faster components.
K   speedup for new component.

*Flynn Benchmark*

Execution time = total time required to run program, that is, product development wall clock time and research [12, 13].

$performance = \frac{1}{(execution\ time)} \leq 1$ is overwhelmed by concurrency of execution, ILP, I/O speed, processor speed, program type, looping, instruction paths available, and system workload.

*Utilization*

Cantrell [11] generates a formula and benchmark to calculate total executed instructions ($\mu$), if T seconds is the mean time of execution, $\mu$ is expressed by $\mu = \frac{1}{T}$

$$The\ Utilization: \rho = \lambda T = \frac{\lambda}{\mu} = 1.$$

When, mean waiting time, Tw = $\infty$, and $\lambda$ = *rate of issue of instructions*. This implies that parallelism is not available in the program.

## 4   Results and Discussion

ILP approach experimental execution time versus predicted approach is shown in Table 2.

Table 2 gives details of the implementation of the mean time of executions in ILP mechanism test for 50 students, 10,000 and over instructions set, 4 sub-loops

**Table 2** Simulation mean time of executions

| Time of execution | Loop fields | | | |
|---|---|---|---|---|
| | 11 | 12 | 13 | 14 |
| T0 (sec) | 952 | 374 | 476 | 604 |
| T1 (sec) | 401 | 137 | 143 | 431 |

forms executed in parallel and simultaneously with predicted mechanism and $\pm 0.05$ statistical sample error students records.

Total execution time, $Tt = \sum T0 + \sum T1$
where

T0   time of execution for predicted technique.
T1   time of execution for ILP technique.

$$\sum T0 = 851 + 205 + 337 + 514 = [1] = 1907$$

$$\sum T1 = 502 + 306 + 282 + 291 = [2] = 1381$$

$$\therefore Tt = \sum T0 + \sum T1 = [1] + [2]$$

$$= 1907 + 1381$$

$$= 3288$$

Predicted and ILP techniques percentages of time of execution are given by

$$Percentage\ of\ T0 = \frac{\sum T0}{\sum Tt} \times \frac{100}{1}$$

$$= \frac{1907}{3288} \times \frac{100}{1} = 0.5799878 \times 100 \quad = 579987 = 58\%$$

and

$$Percentage\ of\ T1 = \frac{\sum T1}{\sum Tt} \times \frac{100}{1}$$

$$= \frac{1381}{3288} \times \frac{100}{1} = 0.4200122 \times 100 \quad = 4200122 = 42\%$$

The performances ($P_0$) of predicted technique and ($P_1$) of ILP are given by

$$P0 = \frac{1}{Te0} = \frac{1}{1907} = 5.2438 \times 10^{-4}(Approx.)$$

$$P1 = \frac{1}{Te1} = \frac{1}{1381} = 7.2411 \times 10^{-4}(Approx.)$$

Units are cycle per second (CPS). The performance rate of ($P1$) improved over ($P0$) due to lesser execution time of the ILP technique. The speedup ($n$) is computed from Eq. 1 by

$$n = \frac{T0}{T1} = \frac{1907}{1381} = 1.381(Approx.)$$

This signifies that the ILP mechanism is 1.381 times faster than predicted mechanism.

*Comparisons*: To estimate the strength and direction of a linear relationship between the outcomes of two approaches for executing branched instructions constructs, linear correlation coefficient ($r$) statistical quantity is used as expressed in Eq. 3:

$$r = \frac{n\sum T0.T1 - (\sum T0)(\sum T1)}{\sqrt{n(\sum T0^2) - (\sum T0)^2}.\sqrt{n(\sum T1^2) - (\sum T1)^2}}. \tag{3}$$

where n is numbers loop iterations $= 4$,

$$\sum T0 = 2406,$$

$$\sum T1 = 882,$$

$$\sum T0.T1 = 622462,$$

$$\left(\sum T0\right)^2 = 1637572,$$

$$\left(\sum T1\right)^2 = 240420,$$

$$n\sum T0.T1 - \left(\sum T0\right)\left(\sum T1\right) == 4 * 622462) - (2406 * 882) = [3] = 367756$$

$$\sqrt{n\left(\sum T0^2\right) - \left(\sum T0\right)^2}.\sqrt{n\left(\sum T1^2\right) - \left(\sum T1\right)^2} = [4]$$

$$= -2480389922$$

$$r = \frac{[3]}{[4]} = \frac{367756}{-2480389922} - 0.000148265$$

The value of linear correlation coefficient ($r$) is (0.000148265), which indicates a negative value tending toward zero. Again, the directions of execution times are inverse of the other (that is, as T0 increases, T1 decreases in long term). The slope of fitness is in the negative region of the graph; and the correlation is random, nonlinear, or weak between the branched instruction execution times of pipeline and TWL techniques.

Similarly, to determine the proportion of the variance (or changes) of predictable variable (T0) from unpredictable variable (T1) for execution times, coefficient of determination ($r^2$) was chosen. This is the ratio of explained variance to total variance given by Eq. 4.

$$r^2 = \left[ \frac{n\sum T0.T1 - (\sum T0)(\sum T1)}{\sqrt{n(\sum T0^2) - (\sum T0)^2} \cdot \sqrt{n(\sum T1^2) - (\sum T1)^2}} \right]^2. \tag{4}$$

where

$$r = -0.000148265,$$

$$r^2 = \left[ \frac{[3]}{[4]} \right]^2 = (-0.000148265)^2 = 0.0000000219826$$

The value of the coefficient of determination ($r^2$) is 0.0000000219826 (or $2.919826 \times 10^8$). This result shows that approximately one-quarter of the area is covered by the line of regression in scatter plot. Therefore, the regression line passes exactly at most one data point in the scatter plot to reveal that the regression line was further away from the points and less explainable. The implication is that ILP technique used lesser execution times as compared to the predicted technique because, it increased numbers of iterations, multiple issues, parallel execution of both conditional and straight instructions constructs, and availability of overlaps in loops processes.

*Users* estimate ILP from execution time simply as overall time needed to run a program loop. According to Flynn's benchmark, the numbers of instructions revolved per cycle is equal or less than 1 ($P1 \leq 1$). That signifies that for improved technique with execution time of 1381; recall that: $P_1 = \frac{1}{time\ of\ execution}$

$$P_1 = \frac{1}{Te1} = \frac{1}{1381} = 7.2411 \times 10^{-4} (Approx.)$$

$P1 \leq 1$ satisfies Flynn's benchmark, i.e., the user will perceive computer system as high-performance system due to parallelism, number of loop pipelines, and I/O speed.

The total work done per unit time is termed *Capacity Index* and can also be perceived as the throughput. The total of requests completed per second is the quantities measured. Utilization is the total instructions issued/total completed per sec.

If average time of execution is $Tt$ seconds, the total instructions executed ($\mu$) is given by

$$\mu = \frac{1}{T0} = 0.00052438 \times 3600 \times 24 = 0.3020451$$

$$\mu = \frac{1}{T1} = 0.00072411 \times 3600 \times 24 = 62.563104$$

and $\lambda = no\ of\ loops = 4$

From Eq. 3, utilization for:
Predicted technique, $\rho = \lambda T0$

$$= \frac{\lambda}{\mu} = \frac{4}{0.3020451} = 13.243056$$

and ILP, $\rho = \lambda T1$

$$= \frac{\lambda}{\mu} = \frac{4}{62.563104} = 0.06393545$$

This implies that $\rho$ for ILP technique is lower than 1 satisfying the Cantrell's benchmark, which means several parallel processes and instructions can be executed in a day.

## 5  Conclusion

The approach revealed that ILP exploits branched executions better than predicted method on the basis of the time of execution, rate of execution, and utilization indexes. ILP offers increased speed of execution in programs, enabling capabilities of processing elements/compiler to point several instructions concurrently.

More so, ILP converts basic block of instructions to more dependent and independent instructions, performs parallel executions unlike the predicted approach (pipeline). ILP makes compiler/processing elements to overlap (or interleave) execution, replicates the original loops (loop iterations) loop copies, thereby encouraging parallel execution of several loop bodies at the same time and several issues of instructions.

This paper recommends ILP approach as the most preferable choice for speedup and parallel execution of instructions from either instruction constructs as compared to pipeline approach. Comparing several techniques with ILP could be considered as future work.

**Acknowledgements**  We acknowledge the support and sponsorship provided by Covenant University through the Centre for Research, Innovation and Discovery (CUCRID).

## References

1. Hennessy, J., Patterson, D.A.: Computer Architecture, 4th edn, pp. 2–104. Morgan Kaufmann Publishers Elsevier, San Francisco (2007)
2. Smith, J.E., Weiss, J.: PowerPC 601 and Alpha 21064: A tale of two RISCs. J. Comput. IEEE Press **27**(6), 46–58 (1994)

3. Jack, W.D., Sanjay, J.: Improving instruction-level parallelism by loop unrolling and dynamic memory disambiguation. Unpublished M.Sc. thesis of Department of Computer Science, Thornton Hall, University of Virginia, Charlottesville, pp. 1−8 (1995)
4. Pepijn, W.: Simdization transformation strategies - polyhedral transformations and cost estimation. Unpublished M.Sc thesis, Department of Computer/Electrical Engineering, Delft University of Technology, Netherlands, pp. 1−77 (2012)
5. Vijay, S.P., Sarita, A.: Code transformations to improve memory parallelism. In: 32nd Annual ACM/IEEE International Symposium on Microarchitecture, pp. 147 − 155. IEEE Computer Society, Haifa (1999)
6. Pozzi, L.: Compilation techniques for exploiting instruction level parallelism, a survey. Department of Electrical and Information, University of Milan, Italy Technical Report 20133, pp. 1−31 (2010)
7. Parthasarathy, K.A.: Performance measures of superscalar processor. Int. J. Eng. Technol. IJET Publications, UK 1(3), 164–168 (2011)
8. Kaeli, D., Rosano, R.A.: Exposing instruction level parallelism in presence of loops. J. Comput. Syst. 8(1), 74–85 (2004)
9. Marcos, R.D.A., David, R.K.: Runtime predictability of loops. In: 4th Annual IEEE International Workshop on Workload Characterization, I.C., Ed., Texas, pp. 91−98 (2001)
10. Misra, S., Alfa, A.A., Adewale, O.S., Akogbe, A.M., Olaniyi, M.O.: A two-way loop algorithm for exploiting instruction-level parallelism in memory system. In: Beniamino, M., Sanjay, M., Ana, M., Rocha, A.C. (eds.) ICCSA 2014, LNCS, vol. 8583, pp. 255–264. Springer, Heidelbreg (2014)
11. Cantrell, C.D.: Computer system performance measurement. Lecture CE/EE 4304, The University of Texas, Dallas (2012). http://www.utdallas.edu/~cantrell/ee4304/perf.pdf
12. Flynn, M.J.: Computer Architecture: Pipelined and Parallel Processor Design, 1st edn, pp. 34–55. Jones and Bartlett Publishers, New York (1995)
13. Olukotun, K., Nayfeh, B.A., Hammond, L., Wilson, K., Chang, K.: The case for a single-chip multiprocessor. Conf. ACM SIGPLAN Not. Stanf. 31(9), 2–11 (1996)

# Design and Implementation of an E-Policing System to Report Crimes in Nigeria

Nicholas Omoregbe, Sanjay Misra, Rytis Maskeliunas,
Robertas Damasevicius, Adesola Falade and Adewole Adewumi

## 1 Introduction

There is no doubt about the importance of the internet; it has made people more connected than ever before. The use of the internet by the police could help the public easily get access to them, thereby making crime reporting easier and faster responses can be gotten, which will make crime fighting a lot easier. As defined by the Mohamed [1], "E-Policing is the exchange of information and services amongst police, related law enforcement agencies and citizens via internet".

There are several areas of an E-policing system, but in this work, we are handling crime reporting in E-policing. Crime reporting involves the creation and submitting of crime reports. Crime reports are written description of crime events, which is gotten from one or more sources (location of crime) upon which decisions will be made. The proposed system will allow a police staff or citizens report crime incidents, accidents, and every other distasteful event, when they come in contact with such. The system allows upload of videos (if available), audio files, images from the crime

N. Omoregbe · S. Misra (✉) · A. Falade · A. Adewumi
Department of Computer and Information Sciences, Covenant University, Ota, Nigeria
e-mail: sanjay.misra@covenantuniversity.edu.ng

N. Omoregbe
e-mail: nicholas.omoregbe@covenantuniversity.edu.ng

A. Falade
e-mail: adesola.falade@covenantuniversity.edu.ng

A. Adewumi
e-mail: wole.adewumi@covenantuniversity.edu.ng

R. Maskeliunas · R. Damasevicius
Kaunas University of Technology, Kaunas, Lithuania
e-mail: rytis.maskeliunas@ktu.lt

R. Damasevicius
e-mail: robertas.damasevicius@ktu.lt

© Springer Nature Singapore Pte Ltd. 2019
R. K. Shukla et al. (eds.), *Data, Engineering and Applications*,
https://doi.org/10.1007/978-981-13-6351-1_21

scene and description of the whole crime event. Crime reports by police personnel are more detailed, depending on the type of crime. It can include major suspects.

Information technology has made people's jobs in various sectors of the industry easier and the same can be said about the Nigerian police force if adopted. E-policing (crime reporting) will allow the police get fast access to incidents occurring in an area, in which there is no police personnel available. Also, a database of crime reports means a report can easily be lookup and decisions can take on them.

## 1.1 Aim, Objective and Significance of Study

The importance and relevance of an E-policing system in Nigeria cannot be overemphasized as the country is still lacking behind in automating some of its key security system such as E-policing and so on. E-policing system is a boost to the efficiency of crime reporting, with the deployment of an E-policing system, crime reporting becomes easy and faster. Citizens can easily report criminal incidence wherever they are, at any time. For the police, the E-policing system in Nigeria provides an organized record of complaints, criminal data, cases and records that can be easily accessed by the authorities (police force). Also for the police, the E-policing system delivers convenient and cost-effective services. The E-policing system provides a platform for affiliations with corporate organizations that are technologically inclined. With the current state of Internet availability through mobile networks and phones, citizens in the villages, remote and rural areas can access the internet and report their cases. Most citizens dread the word "police station", so they are unwilling to go to the police station to write statements, this e-police system makes it easy to report cases without visiting the police station.

## 2 Review of Related Findings/Technologies

Colgan [2] designed a complete E-policing system for community policing. This system includes major features of a typical police activity such as crime reporting, investigation, documentation, crime monitoring and the likes. The system has been able to computerize policing activity in a community but it did not provide a fast and easy means for reporting crimes.

Fagbohun [3] designed a prototype for a crime reporting E-policing system. This system records geographical areas, streets, occupants and such information in the police control centre. This information is required for maximum functionality. The system consists of a one-touch sender, distress point call unit and a local area electronic policing controller designed to monitor and control each station. This system provides a faster transmission time and method for criminal activity.

Information technology is introduced by traditional into their activities to boost their competitive edge, improve performance, facilitate management and also develop new potentials in their work [4]. Other components of a society are not left out. Technological changes are guided by three parallel fundamentals in policing: to enhance efficiency and effectiveness, to conform to the demands of improved forms of police accountability/management and in addition to meet external agencies for information growing demand.

## 2.1 Efficiency and Effectiveness

First, efficiency and effectiveness imperative is technology-inclined. Technology and police work has always been closely related. Technology does not only enhance police efficiency and effectiveness in eradicating crime but it also enhances their professionalism in executing their duties effectively.

Manning [5] and Davids [6] explained that the stock-in-trade of policing is information; naturally, the latest information technologies would be embraced by police organizations. Information technology is being invested by the police to raise their capacity to process and store appreciable amount of data; to provide full accessibility to criminal records and other information that is crime-related; and to improve their intelligence and investigative capabilities.

## 2.2 New Public Management and Accountability in Policing

Second, public management and accountability imperative is policy-inclined. Police organizations and commercial firms are quite similar in the sense that they both make use of information technology to improve performance. Since 1980s, a new era of public accountability has increased in major Western democracies such as Britain and Australia [7–10].

Traditionally, police procedures and practices are coordinated by rules and laws that are compelled by police hierarchies and the courts, respectively. Deterrence through investigation and enforcement, legislation and rule-making, criminal sanctions and organizational discipline is the major mode of control. Power [8], Miller and Nikolas [9] and O'Malley and Palmer [11] explained the recent accountability for public organizations, the uses of techniques at managerial level and private administrative structures for corporation's interest, efficiency, emphasizing cost control and decentralization of management, whilst creating quasi-market mechanisms such as risk assessment and audit procedures, contracting out, performance indicators. Power [8] and Mullen [12] traditional police force has changed to organizations with a new emphasis on crime management and marketing strategies, mission statements, business plans, customer service and performance measures due to the recent managerialism in policing.

## 2.3  External Demands for Information

The demand from external agencies for information such as information crime and accident data for external bodies such as insurance companies and road traffic authorities is relatively high. This information is given at regular interval by the police organizations and watchdog agencies demanding for records regarding police accountability and actions. Therefore, there is a need to improve information technology capacities within police organizations for external demands for police records.

## 2.4  Information Technology on Policing and Its Effects

Marcel-Eugene [13] explained the pervasive usage of IT, and it plays a major role in everyday life for many individuals and quite a number of organizations. But how visible this adoption is to be realized when it comes to the police force is an important aspect to consider [5]. Through the internet, our channels of communication are expanded by E-policing but it does not replace face-to-face contact or telephone, which is extremely important. Mell and Grance [14] developed an E-policing initiative which requires the development of a strategic vision and direction, evaluation of technology options, evaluating impacts of related legislation, predicting costs and recognizing limitations and resistance.

Improved approach of performing duties is open for the police through E-policing, improved tools and interactive flow of information between the citizen and the police. E-policing enables a dual communication and improved accessibility for police and public. Mell and Grance [14] stated that E-policing shows that police are in tune with present technology.

## 3  Context Diagram and Conceptual Framework of E-Policing

Figure 1a indicates a context diagram for implementing an E-policing system in Nigeria whilst Fig. 1b indicates a regular collaborative virtual E-policing reference framework for implementing an integrated solution based on internet technology amongst locations in Nigeria.

**(a)**   **(b)**

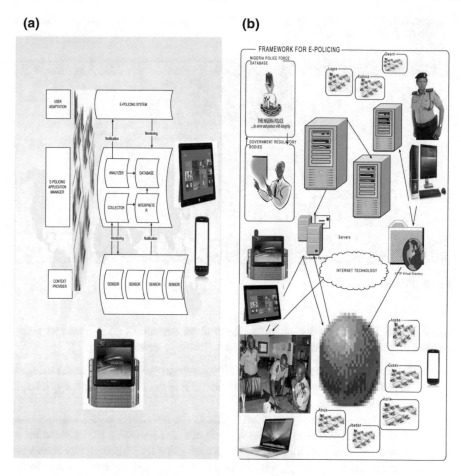

**Fig. 1**  **a** Context layout. **b** Conceptual framework

## 3.1  The Context Layout

The context layout focused on notification usage and monitoring systems that allow for the provision of several services, especially user adaptation services and mobile communication services.

## 3.2  The Conceptual Framework

Conceptual framework focused on the usage of connected police services, which allow the provision of several services especially communication services. Aggre-

**Fig. 2** The home page

gated E-policing framework indicates the mutual communication amongst stake-holders and infrastructures for information technology involved (Fig. 2).

- Regulatory bodies for government and NPF database certify proper authentication and accreditation of registered police officers. It also enables efficiency, reliable recording and database security.
- Different remote locations discussed earlier in the introduction and abstract indicates that police officers and police stations can communicate in real time heedless of their geographical locations. People requesting for police services can have access to any station through any of the internet-enabled device attached to the central server which will be located in the cloud as shown in Fig. 3.

## 4 Methodology

The study used programming tools such as JAVA script, CSS, HTML, Dreamweaver and PHP. These languages were selected due to their platform in-dependability, the wide acceptability they enjoy and open-source nature. The back-end design was designed using MySQL; a useful tool in designing relational databases. All these are resident on an Apache-hosted web server (the middleware). To address the security issue, passwords have also been introduced to allow access to records on role-based only (Figs. 4 and 5).

**Fig. 3** The contact us page

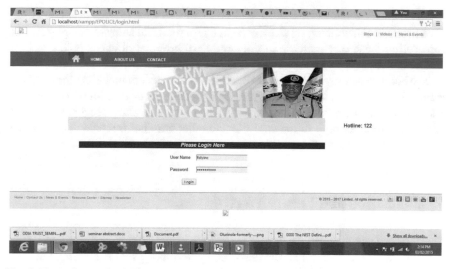

**Fig. 4** The police station login page

## 4.1 Presentation and Validation of the Model

Just as continuity of safety of Nigerian citizens requires a cooperative environment, the development of a system using modern technologies that aid internet technology in policing services is an increasing necessity [15]. Hence, provision of a secure and easy to implement environment for E-policing applications will give notable positive effect in present computer systems and networks. It is highly necessary to increase

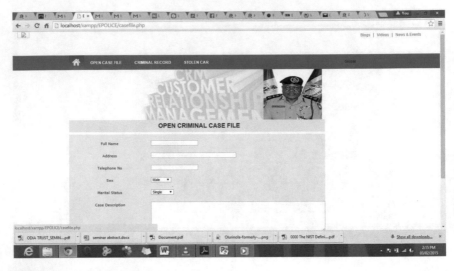

**Fig. 5** The criminal case form

access to essential policing services, especially for rural and underserved populations considering the steady increase in the number of criminal offences police stations receive daily.

The home page provides a general overview of the system including the available menus. The contact us form enables the general public to report crimes and give necessary advice and information to the NPF. The login page allows registered officers or police station to login to the case file and other criminal offences. The criminal case form allows criminals or suspects to fill the case form electronically.

## 5  Conclusion and Future Work

This research has analysed the urge to improve information flow and communication amongst stakeholders in NPF. The designed system shows that E-Policing will help in eradicating the barriers of manual policing system in Nigeria.

Performing the task of maintainability which is a process usually comes with a heavy cost in any software development. Software life span increases with maintenance. Future works are to show a detailed statistical analysis. Preventing unauthorized users from gaining access into the system, a more sophisticated security measure will also incorporate into the implementation of the system. The evaluation of E-Policing in Nigeria will be done by benchmarking it with the existing systems in the USA and other developed countries.

**Acknowledgements** The authors gratefully acknowledge the support of Department of Computer and Information sciences, Covenant University and the NPF for access to information and guidance towards the success of the work.

# References

1. Mohamed, T.: A proposed model-based security framework for E-policing. In: 2011, IEEE International Conference Computer Applications and Network Security (ICCANS 2011), IEEE (2011)
2. Colgan, G.: Electronic apparatus for implementing community policing program and method therefor. U.S. Patent 5,510,978 (1996)
3. Fagbohun, O.: Improving the policing system in Nigeria using electronic policing. J. Eng. Appl. Sci. 2(7), 1223–1228 (2007)
4. Earl, J.: Experiences in strategic information systems planning. MIS quarterly 1–24 (1993)
5. Manning: Information Technologies and the Police, Tonry, M., Morris, N. (eds.) (1992)
6. Davids, C., Hancock, L.: Policing, Accountability, and Citizenship in the Market (1998)
7. Chan, J.: Changing Police Culture: Policing in a Multicultural Society. Cambridge University Press, Melbourne (1997)
8. Power, M.: The Audit Society: Rituals of Verification. Oxford University Press, Oxford (1997)
9. Miller, P., Nikolas, R.: Governing Economic Life: Economy and Society, vol. 19, no. 1, pp. 1–31 (1990)
10. Yu, Y., Alison, D.: The contribution of emotional satisfaction to consumer loyalty. Int. J. Serv. Ind. Manag. 12(3), 234–250 (2001)
11. O'Malley, P., Palmer, D.: Post-Keynesian Policing Econ. Soc. 25(2) (1996)
12. Mullen, K.L.: The Computerization of Law Enforcement: A Diffusion (1996)
13. Marcel-Eugene, L:. E-Policing in Police Services—Definitions, Issues and Current Experiences (2006)
14. Mell, P., Grance, T.: The NIST definition of cloud computing. Nat. Inst. Stand. Technol. 53(6), 1–7 (2009)
15. Wiki Educator, Police Crime Report, site: http://wikieducator.org/Lesson_2:_police_crime_report. Accessed 17 Jan 2014

# Performance Evaluation of Ensemble Learners on Smartphone Sensor Generated Human Activity Data Set

Dilip Singh Sisodia and Ankit Kumar Yogi

# 1 Introduction

The smartphones are equipped with powerful preinstalled sensors such as audio sensors, GPS sensors, image sensors, temperature sensors, light sensors, acceleration sensors, and direction sensors which are very much useful for a human [1]. These small mobile devices with fast computing powers to send and receive data create a whole new domain for data mining applications and research.

All the smartphones and music players, including the iPod and the iPhone [2], have triaxial accelerometers installed, which measure the acceleration in all the three spatial dimensions. These accelerometers are highly capable of detecting the location of the mobile device which is very useful and important information for activity recognition. Initially, the accelerometer was installed into these devices to enable the screen rotation and to support advanced game playing, but later the use extended to recognize a user's activity and many other useful applications. The phone can gather the information about the activity which a human performs and can be used by various applications as per the requirement.

Activity recognition of a human's body is not a new concept [3]; the concept is used to recognize almost 25 human body activities using the biaxial accelerometer situated in five different locations on the human body. The data used to predict user's energy consumption [4], the movement of the user after the fall [5], and user's activity level so that it can be used to promote user's health and fitness [6]. Earlier studies involved smartphones that focused on a very small set of users [7] and for particular

D. S. Sisodia (✉)
National Institute of Technology Raipur, Raipur, India
e-mail: dssisodia.cs@nitrr.ac.in

A. K. Yogi
Jaypee University of Engineering & Technology, Guna, India
e-mail: yogiankit73@gmail.com

© Springer Nature Singapore Pte Ltd. 2019
R. K. Shukla et al. (eds.), *Data, Engineering and Applications*,
https://doi.org/10.1007/978-981-13-6351-1_22

user [8] they had trained models rather than making a universal model which can be applied to any user.

This paper discussed the use of accelerometer sensor installed on a mobile phone, to recognize human activity. The data is taken from mobile phones based on the Android operating system because such phones are easy to handle, operate and are also becoming popular day by day covering the whole market.

The ensemble learning algorithms, namely, random forest, AdaBoost, and bagging are used on the data generated by more than 29 users while performing various physical activities such as jogging, sitting, standing, walking, and climbing stairs. The performance of used ensemble learners is evaluated using accuracy, f1 score, recall, and precision.

Rest of the paper is organized as follows: Sect. 2 discusses related work, and Sect. 3 defines the process, which addresses the activity recognition task and collection of data, preprocessing of data and transformation of data. Section 4 gives us a brief knowledge about the experiments, and the results and Sect. 5 summarizes the future work along with the conclusion.

## 2   Related Work

The availability of accelerometers in smartphones is increasing day by day, capturing of human activity is becoming very easy. Enormous work has already been reported on the activity recognition using several accelerometers positioned on different body parts to get the of the user's activity record. In [3] data recorded to track the change in user's daily activity, five accelerometers on 20 user's are analyzed using C4.5 and Naive Bayes classifiers and made a model capable of recognizing 20 activities which a user can perform. The results demonstrated that the accelerometers which were placed at thigh proved to be more powerful in recognizing the user's activity as compared to the others placed on different body parts.

In [9], two accelerometers are employed on three subjects for recording five human activities such as to sit, to walk, to run, to stand, and lie down. The results of this paper claim that the thigh accelerometer was unable to detect activities properly. Therefore, a need arises to place accelerometers at various body parts to resolve the issue. In [10], the data collected from ten different users wearing accelerometers on their lower body parts is tested and analyzed by applying different learning methods. In [11], activity data recorded by putting five accelerometers at different body locations of 21 subjects. The data is used for implementation of a real-time model for recognizing 30 activities added the dataset from the heart monitor along with the accelerometer data.

In [12], authors utilized five triaxial accelerometers attached to different body locations such as wrist, ankle, hip, thigh, and arm to identify 20 activities of 13 different users. The authors used several learning methods to recognize various postures such as standing, lying, sitting, and five movements which include walking, stair climbing, running, and cycling. In [13], five accelerometers in one experiment to generate the dataset are utilized, out of which two accelerometers were used for activity recognition. The experimentation was performed on 31 males, and a hierarchical classification model was designed to easily differentiate between the postures like lying and sitting at different specific angles and also to distinguish between the motions of walking and climbing stairs at different speeds.

In [14], authors identify the user activities such as standing, lying, walking, running, swimming, football, playing ball, croquet, and for using the toilet in the specific locations of subjects by 20 different types of sensors worn on wrist and chest. In [15], activities and locations such as standing, sitting, walking on the ground, walking downstairs, and walking upstairs using the biaxial accelerometer in the sensor module and also used angular velocity sensor kept in the pocket with the digital compass worn by the user at the waist are recognized. In [16], a model capable of recognizing same activities was built, and the dataset was taken from the triaxial accelerometer, phototransistors, temperature and barometric pressure sensors, two microphones, and GPS to distinguish between a stationary state, walking, jogging, driving a vehicle, and climbing up and down stairs. The other systems were not very practical as they involved multiple sensors situated all across the body but in [17] author used a system which involved various accelerometers or combination of accelerometers capable of identifying a wide range of activities. The system can work for only some small-scale applications (e.g., hospital setting). In [18], six activities are identified by employing a watch fitted on the belt, shirt's pocket, backpack, and trouser pocket. The "e-Watch" dwelled with the biaxial accelerometer and the light sensor and used four classifiers along with five-fold cross-validation. In [4], seven different types of sensors are utilized for the recognition of activities such as walking, sitting, standing, ascending stairs, descending stairs, elevator moving up and down, and brushing one's teeth.

## 3   Activity Recognition Task

This section describes the processes for activity recognition and also describes activity recognition. Section 3.1 tells about the raw accelerometer data that have been collected, Sect. 3.2 describes the processing and transformation of raw data into a set of examples and describes the activities that are to be identified or are going to be identified.

## 3.1   Collection of Data

To collect data for supervised learning, it is mandatory to have large Android-based smartphone users who carry their smartphone with them while they are performing an activity. The data generated by different users were evaluated who carried their smartphone while performing any task like they carried their smartphone in the front pants leg pocket while they were standing, sitting, walking, running, jogging, ascending stairs or descending stairs for a particular period.

The collected data is controlled by a phone-based application which is designed to fulfill the same. This application helps to keep the record of the user's name, the start and stop time of the user and the activity completed by the user.

## 3.2  Materials and Methods

The raw time data should be transformed at first as Standard classification algorithm because it cannot be applied directly. Therefore, the data is first preprocessed and transformed into training examples [9]. In the next step, the training dataset is used to make a model to recognize the activities such as jogging sitting, standing, walking, and climbing up and down stairs. Using the information or the dataset, the model recognizes the activities to keep a record of them, which can be used to help the user evaluate whether his daily activity is performing, and the accelerometer also plots the graph of the activities a user performs [10].

## 4  Experiments and Results

In experimentation, the raw data is collected from WISDM. The WISDM is an Android-based data collection platform consists of software and hardware architecture used to channelize the human activity data through a sensor-based smartphone to the Internet-based server and then the data is transformed. The steps are taken to evaluate the results are shown in Fig. 1, the first step is to take the raw data set which includes the missing values in multi-class than the average of the values are calculated, and the missing values are replaced with the average value. In the next step, the three classifiers are applied at ten-fold cross-validation, and the accuracy of all the three classifiers is evaluated. Moreover, a model is designed which predicts the daily activities such as jogging, sitting, standing, walking, climbing stairs, down-stairs with the help of a cell phone based accelerometer that allows a user to have a daily check on his routine whether he is performing his daily routine regularly or not.

The experiment uses a dataset and records the accuracy from three different classifiers, i.e., random forest, AdaBoost, and begging for ten k-folds, and then the average accuracy is calculated. The random forest classifier results in maximum accuracy, and AdaBoost classifier results in the least accuracy.

The results as shown in Table 1 details about the various parameter values obtained on applying different classifiers on the dataset. Random forest classifier outperforms the other two on every parametric value and results into the maximum accuracy of 90.20% which is very high compared to AdaBoost classifier which yields only 67.83% accuracy. The accuracy of bagging classifier is 89.27% which is very much

**Fig. 1** Process flow of human activity recognition

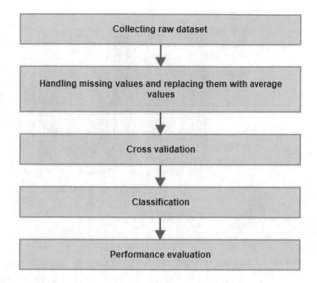

**Table 1** Percentage of record predicted correctly

| Classifier | Avg. accuracy (%) | F1 score (%) | Recall (%) | Precision (%) |
|---|---|---|---|---|
| Random forest | **90.20** | 90.02 | 90.20 | 90.17 |
| AdaBoost | 67.83 | 61.84 | 67.83 | 61.10 |
| Bagging | 89.27 | 89.36 | 89.27 | 89.76 |

close to that of random forest classifier with 90% accuracy, but the best average result is obtained from random forest classifier.

On the parameter of the F1 score, AdaBoost classifier gives least performance of 61.84% followed by the Bagging classifier with 89.36%. Random forest classifier results into 90.02% F1 score and 90.20% recall value followed by the Bagging with 89.27% and AdaBoost with 67.83% recall value. The precision value for Bagging classifier is 89.76% almost equal to that of Random forest classifier with 90.17%. The AdaBoost again results into least precision of 61.10%.

The performance of AdaBoost classifier on all parametric values is very close to the random forest. However, the best average results are obtained using the random forest. Therefore, out of the three classifiers used Random forest gives the best average result.

The graph in Fig. 2 distinguishes between the average accuracy, the f1 score, recall, and the precision obtained using the three classifiers, and it can be seen that the average best results are obtained from the random forest classifiers.

A confusion matrix is a way of classifying true positives, true negatives, and false positive and false negatives when there are more than two classes. Confusion matrix makes it easy to compute the precision and recall of a class. It is used for computing

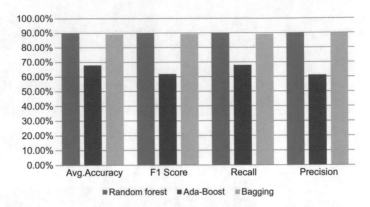

**Fig. 2** Performance of different Classifiers

the precision and recall and hence f1 score for multi-class problems. The columns in Table 2 represent the actual values, and rows represent the predicted values.

In Tables 2, 3, and 4 the confusion matrices are presented which are associated with the three classification algorithms. The confusion matrix indicates the prediction and prediction errors due to confusion between the activities.

Table 2 gives the confusion matrix for random forest classifier in which jogging class is predicted correctly.

**Table 2** Confusion matrix for random forest classifier

| | | Predicted class | | | | | |
|---|---|---|---|---|---|---|---|
| | | Walk | Jog | Up | Down | Sit | Stand |
| Actual class | Walk | **41** | 0 | 0 | 1 | 10 | 10 |
| | Jog | 1 | **176** | 0 | 0 | 3 | 1 |
| | Up | 0 | 0 | **24** | 0 | 0 | 0 |
| | Down | 0 | 0 | 2 | **25** | 0 | 0 |
| | Sit | 6 | 2 | 0 | 0 | **33** | 9 |
| | Stand | 1 | 1 | 0 | 0 | 6 | **189** |

**Table 3** Confusion matrix for AdaBoost Classifier

| | | Predicted class | | | | | |
|---|---|---|---|---|---|---|---|
| | | Walk | Jog | Up | Down | Sit | Stand |
| Actual class | Walk | **0** | 3 | 0 | 0 | 4 | 50 |
| | Jog | 0 | **136** | 0 | 0 | 3 | 28 |
| | Up | 0 | 0 | **19** | 7 | 0 | 0 |
| | Down | 1 | 1 | 8 | **12** | 0 | 1 |
| | Sit | 0 | 7 | 0 | 2 | **8** | 47 |
| | Stand | 0 | 6 | 0 | 1 | 3 | **192** |

**Table 4** Confusion Matrix
for Bagging Classifier

| | | Predicted class | | | | | |
|---|---|---|---|---|---|---|---|
| | | Walk | Jog | Up | Down | Sit | Stand |
| Actual class | Walk | **28** | 1 | 0 | 1 | 6 | 7 |
| | Jog | 2 | **157** | 0 | 0 | 4 | 2 |
| | Up | 0 | 0 | **25** | 0 | 0 | 0 |
| | Down | 1 | 0 | 1 | **23** | 1 | 0 |
| | Sit | 12 | 2 | 0 | 0 | **38** | 8 |
| | Stand | 10 | 0 | 0 | 0 | 0 | **212** |

Table 3 gives us the confusion matrix about the AdaBoost classifier.

Table 4 gives us the confusion matrix about the bagging classifier.

The diagonal elements represent the number of points for which the predicted label is equal to the true label, while off-diagonal elements represent those that are mislabelled by the classifier. The higher the diagonal values of the confusion matrix, the better, indicating many correct predictions.

# 5 Conclusion

This paper compares the performance of ensemble learners on human activity dataset obtained from WISDM Android-based data collection platform. The WISDM keeps a record of the basic human activities such as jogging, sitting, standing, walking, climbing stairs, downstairs in the day to day life. The raw dataset is preprocessed and prepared for experimentation. The experiments are performed on human activity dataset and record the performance from three different classifiers, i.e., random forest, AdaBoost and begging for ten k-folds, and then the average accuracy is calculated. The classifiers performance is evaluated on the basis of accuracy, average accuracy, f1 score, recall, and precision are calculated along with the confusion matrix. The random forest classifier results in maximum accuracy, and AdaBoost classifier results in the least accuracy.

In future, experiments may be performed with added activities, for instance, bicycling and car driving. The experimental data may be retrieved from users for the possible improvement in the results. More refined and practical features can be developed when collecting the raw time-series data. The impact of placing the cell phone at user's different body locations like as on a belt loop may be tested.

# References

1. Weiss, G.M.: WISDM (Wireless Sensor Data Mining) Project. Fordham University, Department of Computer and Info. Science (2011)
2. Nolan, M., Mitchell, J.R., Doyle-Baker, P.K.: Validity of the Apple iPhone{®}/iPod Touch{®} as an accelerometer-based physical activity monitor: a proof-of-concept study. J. Phys. Act. Health **11**, 759–769 (2014)
3. Bao, L., Intille, S.: Activity recognition from user-annotated acceleration data. Pervasive Comput. 1–17 (2004)
4. Choudhury, T., Hightower, J., Rahimi, A., Rea, A., Hemingway, B., Koscher, K., Landay, J.A., Lester, J., Wyatt, D.: The mobile sensing platform: an embedded activity recognition system. IEEE Pervasive Comput. **7**, 32–41 (2008)
5. Mathie, M.J., Lovell, N., Sydney, U.: Classification of basic daily movements using a triaxial accelerometer. Med. Biol. Eng. Comput. **42**, 679–687 (2004)
6. Anderson, I., Maitland, J., Sherwood, S., Barkhuus, L., Chalmers, M., Hall, M., Brown, B., Muller, H.: Shakra: tracking and sharing daily activity levels with unaugmented mobile phones. Mob. Netw. Appl. **12**, 185–199 (2007)
7. Yang, J.: Toward physical activity diary: motion recognition using simple acceleration features with mobile phones. In: Proceedings of the 1st International Workshop on Interactive Multimedia for Consumer Electronics, pp. 1–10 (2009)
8. Brezmes, T., Gorricho, J.L., Cotrina, J.: Activity recognition from accelerometer data on a mobile phone. In: Distributed Computing, Artificial Intelligence, Bioinformatics, Soft Computing, and Ambient Assisted Living, pp. 796–799 (2009)
9. Weiss, G.M., Hirsh, H.: Learning to predict extremely rare events. In: AAAI Workshop on Learning from Imbalanced Data Sets, pp. 64–68 (2000)
10. Kwapisz, J.R., Weiss, G.M., Moore, S.A.: Activity recognition using cell phone accelerometers. ACM SIGKDD Explor. Newsl. **12**, 74–82 (2011)
11. Krishnan, N.C., Colbry, D., Juillard, C., Panchanathan, S.: Real time human activity recognition using tri-axial accelerometers. In: Sensors, Signals and Information Processing Workshop, pp. 3337–3340 (2008)
12. Tapia, E.M., Intille, S.S., Haskell, W., Larson, K.W.J., King, A., Friedman, R.: Real-time recognition of physical activities and their intensities using wireless accelerometers and a heart monitor. In: International Symposium on Wearable Computers, pp. 37–40 (2007)
13. Mannini, A., Sabatini, A.M.: Machine learning methods for classifying human physical activity from on-body accelerometers. Sensors **10**, 1154–1175 (2010)
14. Foerster, F., Fahrenberg, J.: Motion pattern and posture: correctly assessed by calibrated accelerometers. Behav. Res. Methods **32**, 450–457 (2000)
15. Juha, P., Peltola, J.: Activity classification using realistic data from wearable sensors. IEEE Trans. Inf Technol. Biomed. **10**, 119–128 (2006)
16. Lee, S.W., Mase, K.: Activity and location recognition using wearable sensors. IEEE Pervasive Comput. **1**, 24–32 (2002)
17. Subramanya, A., Raj, A., Bilmes, J.A., Fox, D.: Recognizing activities and spatial context using wearable sensors. arXiv preprint arXiv:1206.6869 (2012)
18. Maurer, U., Smailagic, A., Siewiorek, D.P., Deisher, M.: Activity recognition and monitoring using multiple sensors on different body positions. In: International Workshop on Wearable and Implantable Body Sensor Networks, BSN 2006, 4–pp (2006)

# An Insight into Time Synchronization Algorithms in IoT

Neha Dalwadi and Mamta Padole

## 1 Introduction

"The Internet of Things (IoT) is a fact of interrelated things for example computing devices, machines, objects or people that are presented with unique identifiers and ability to transmit data with no interactions required among the things" [1]. Examples of such systems are smart parking, smart appliances, smart metering, smart grid infrastructure, etc. International Telecommunication Union (ITU) and Internet of Things European Research Cluster (IERC) has defined IoT as a "Network infrastructure which is dynamic, global and having self-configuration capabilities created using standard and interoperable communication protocols, where physical and virtual 'things' have identities, physical characteristics and implicit personalities and use intelligent interfaces that get flawlessly incorporated into the information network" [2].

Time synchronization in sensor network becomes significant because in each phase (i.e. management, planning, security) of the network decisiveness is involved when an event occurred. Protocols like Network Time Protocol (NTP) and Simple Network Time Protocol (SNTP) are used as traditional time synchronization protocols. As IoT comprises heterogeneous devices and interfaces, traditional time synchronization protocols are not applicable. As per the traditional approach, time synchronization is performed based on hardware clocks of computers. Similarly, clock synchronization in IoT devices is performing based on the hardware clock of each device. Due to variations in oscillators generate clock drift, hardware clocks are not ideal. Therefore, the time stamp of events will not be observed the same

N. Dalwadi (✉)
Vadodara Institute of Engineering, Vadodara, India
e-mail: neha.dalwadi@gmail.com

M. Padole
The Maharaja Sayajirao University of Baroda, Vadodara, India
e-mail: mpadole29@rediffmail.com

© Springer Nature Singapore Pte Ltd. 2019
R. K. Shukla et al. (eds.), *Data, Engineering and Applications*,
https://doi.org/10.1007/978-981-13-6351-1_23

among the devices in the network. The clock in devices help in identifying, when an event has occurred, for example a sensor measurement is going to be shared in a network. If a device clock is out of synchronization, i.e. it is not aligned with clocks of other devices in the network, then there is a possibility that it may miss messages, collide with other messages being sent by other devices or waste energy in trying to get back in synchronization [3]. Therefore, to keep the network running efficiently, clocks need to be synchronized in order to make data flow and resource availability in a reliable and accurate way. There are many IoT applications in which time synchronization play an important role. Some of the examples are as follows:

a. In sensor network, to save energy of sensor nodes time synchronization is used; it will allow the nodes to sleep for a given time and then awaken periodically to receive a bonfire signal.
b. In smart grid, different parts of electrical grid to connect and disconnect without disrupting customers and it allows networks to impose order on streams of data from scattered sensors.
c. In mobile communications, the sequencing of calls is important when two or more callers trying to communicate with the same receiver.
d. In smart parking systems, the availability or non-availability of parking slot should be immediately and correctly communicated, without causing any collision, to the multiple drivers, who are trying to reach a particular parking slot.

Synchronization in IoT can be achieved by minimizing the clock drift and delay in message passing among the nodes in the network. The paper presents algorithms for time synchronization which are used for wireless sensor networks that can be applied in IoT. These algorithms are based on time communication protocol to approximate reference clock value and sort out deviation in delay in message passing. During this transmission of messages, some factors like a communication link failure, fault tolerance, propagation time, non-receipt of acknowledgment, congestion in a network, the bandwidth of the communication link and routing mechanism may affect and it may raise communication delay during this message passing which directly affects clock synchronization. Before going into detail about synchronization algorithms, it is required to analyze issues in IoT that may affect the performance of synchronization algorithms.

## 2   Issues in IoT

There are five key issues in IoT, these include security, privacy, interoperability, storage and limited energy [1, 4]. These issues directly or indirectly may increase due to time synchronization issue among nodes. For example, each node has limited energy. To minimize energy consumption, nodes need to be in sleep mode for a given time and then stimulate from time to time which requires synchronization between nodes in terms of time in the network.

a. *Security*: As in IoT, devices are integrated. Therefore, the security may be get affected by a weakly secured device which is attached online.
b. *Privacy*: IoT is a wealth of information about those who use it. IoT implementations can revolutionize the behaviour of private data collected, used, analyzed, and protected. Therefore, there is a need to develop the privacy of data in IoT applications.
c. *Interoperability*: Interoperability is required to make seamless connectivity between connected devices, regardless of model, its manufacturer etc.
d. *Storage*: As the storage of data increases, the energy demand of smart devices increases. As a result, there is a need to identify what variety of information would be required to be stored and for what time period.
e. *Limited Energy*: Sensor nodes have limited energy, and IoT system is formed of the wireless sensor network. Therefore, IoT power will degrade due to wasted energy and other environmental effects on sensor nodes.

Time synchronization is one of the central issues among all issues discussed here. For example, in terms of security, IoT technologies with weak synchronization, it takes longer to exchange keys and more vulnerable to unwanted discovery and spoofing. Fast synchronizing protocols are better able to support authentication and can remain in a quiet mode that also protects privacy. IoT requires a new paradigm of combining data and time. There is a need to design synchronization algorithms for time correctness independent of hardware and need determinism and security in networks.

## 3 Time Synchronization in IoT

We have discussed the importance of time synchronization and its issues in IoT applications. In IoT, there is a need for time correction at every node because the physical clock of all nodes may become inaccurate for the following reasons [5]:

a. Randomly switched nodes in the network at arbitrary times.
b. Random frequency oscillation leads to clock drift.
c. Oscillator frequency depends on time and environments.

In IoT, to enhance the utility and importance of sensor data, it is essential that sensors be in synchronized form. But due to sensor heterogeneity, it is difficult to make efficient synchronization algorithms. By achieving time synchronization among nodes, it will also improve data synchronization among the nodes [6].

### 3.1 Time Synchronization Algorithms

Time synchronization algorithms focus on how to minimize delay in message passing and clock skew between nodes in a wireless sensor network. As discussed previously, during transmission of messages many factors affect synchronization between nodes. These factors need to be considered before correcting actual clock value. In the case of wireless sensor network, other factors like energy costs and memory requirement can also affect the performance of time synchronization algorithms. Based on the approach, time synchronization algorithms have the following characteristics [7, 8]:

a. External Clock Synchronization: External real time clock is used as a reference clock for other nodes to synchronize in the network. UTC is used as reference clock time for other nodes in the system.
b. Internal Clock Synchronization: There is no such external clock for time reference. Nodes must agree on the common time of another node in the system.
c. Pair-wise Synchronization: At least one message should be passed between two nodes for synchronization.
d. Network-wide Synchronization: It is the recurrence of pair-wise synchronization all over network.
e. Unidirectional Message Exchange: The time stamp contained in a particular message, is used for synchronization.
f. Two-way Message Exchange: It is based on the assumption of identical propagation delay and clock drift. A receiver node replies with message enclosing three time stamps.
g. Sender–Receiver Synchronization: Sender and receiver clock synchronize with each other using the time stamp value passed by message transmission.
h. Receiver–Receiver Synchronization: The same message is passed to each receiver and synchronization is performed based on message arrival time. Each receiver exchanges their message arrival time to calculate offset value among them.

Following is the description of a set of time synchronization algorithms based on the above characteristics:

1. *Reference Broadcast Synchronization Algorithm (RBS)*:

Receiver–receiver synchronization approach is used by RBS. RBS can be implemented for both types of synchronization internal as well as external. It works as follows [5, 9]:

a. Single-node broadcast a synchronization message to multiple nodes using the network's physical layer broadcast.
b. Receiver nodes record their local time and then exchange this information amongst each other. Referenced broadcast's arrival time is used as a position of reference in favor of evaluating node's clock.
c. Nodes synchronize the arrival time stamp amongst themselves.

RBS uses least-square linear regression method to remove synchronization errors like phase offset and clock skew. The causes of errors in synchronization are message sending time, time required to receive message, transmission time of message and access time. It tries to remove the synchronization errors instead of estimating it. To estimate phase offset and clock skew:

a. Node broadcasts m reference packets.
b. At which time stamp reference was received is noted by all receiver nodes as per their local clock time.
c. Each receiver exchanges this information among them. Phase offset can be computed by a receiver node $N_a$ to receiver node $N_b$ under an assumption is that drift can be abandoned when observations are exchanged.

$T_{Na,Nb}$ = Time stamp of $N_a$'s clock when it received broadcast message from $N_b$.

Phase Offset $[N_a, N_b] = (1/m) * \sum_{(k=1\ \text{to}\ m)} (T_{Nb,k} - T_{Na,k})$.

As we know, clock drift cannot be neglected. Therefore, instead of taking an average of phase offsets from multiple observations, the least-squares linear regression method is performed by RBS. Clock skew is defined as the slope of the best-fit line. Clock correction is not required in RBS. And it maintains the limited energy usage of each node so it is high energy efficient algorithm. The higher computational complexity causes more time to synchronize the clocks in RBS.

Limitation of this algorithm is that the network should support broadcast facility. There is one drawback, a compromised node can exchange an incorrect time and hence an inaccurate offset can be computed by the former node. Collision is a problem because the reference messages prompts all nodes concurrently to tell about their observations. The advantage of using this algorithm is it can be used without external reference clock and there are loosely coupled sender and network interface.

## 2. *Lightweight Tree Based Synchronization Algorithm (LTS)*:

LTS supports both internal and external clock synchronization approaches. There is one referenced clock, which is used by all sensor nodes to synchronize their internal clocks. It does not try to spot on diverse drift rates. It provides a specified precision with a little overhead [5, 10]. There are two approaches:

a. *Pair-wise Synchronization*: It is based on sender–receiver approach.

Suppose node $N_a$ wants to synchronize with node $N_b$ clock. Node $N_a$ triggers 'resync' message at time $t_0$, and at time stamp $t_1$ message is formatted ($L_a (t_1)$) and sends it to node $N_b$. With the consideration of propagation time and reception of message, time assumes that node $N_b$ time stamps received packet at $t_5$ with $L_b (t_5)$. Node $N_b$ replies at time stamp $t_6$ and includes $L_a (t_1)$, $L_b (t_5)$ and $L_b (t_6)$ in reply message. Assume that node $N_a$ receives a reply at time stamp $t_7$ and afterward $N_a$ time stamps it with $L_a (t_8)$. After time $t_8$ node a ($N_a$) processes these four values $L_a (t_1)$, $L_b (t_5)$, $L_b (t_6)$ and $L_a (t_8)$. These time stamp values are used to calculate its clock offset with respect to node $N_b$. It is assumed that in both the directions (request and reply) propagation delay ($\tau$) is the same. Let $t_p$ is the duration for request and answer packets. Both the parameters, that is, propagation delay and duration are assumed to be known to node $N_a$. Node $N_a$ can estimate the clock offset with respect to $N_b$ derived from the recognized values $L_a (t_1)$, $L_b (t_5)$, $L_b (t_6)$ and $L_a (t_8)$. This swap over gets two packets:

1. If node $N_b$ should also determine regarding the offset, one more packet is required from $N_a$ to $N_b$ hauling offset ($T_O$).
2. An improbability is in the interval (I) is estimated as follows: $I = [L_a (t_1) * \tau + t_p, L_a (t_8) - \tau - t_p - (L_b (t_6) - L_b (t_5)]$
3. By choosing the middle of the interval as $Li(t_5)$, the maximum improbability is $|I|/2$ [9, 10].

b. *Network-wide Synchronization*: There is one reference node used as root node. Minimum-height spanning tree is constructed with respect to the reference node. All other nodes in a network synchronize their clock with respect to the reference node. Spanning tree created as follows: node of Level-1 can directly synchronize with reference node. Nodes at Level-2 synchronize with nodes at level-1 and so on. There is one assumption that all synchronization inaccuracies are independent. Hence, synchronization errors are minimized using a minimum spanning.

One drawback with the LTS algorithm is that it relies on the dependability and precision of information from all nodes along the path to the root node. Hence the synchronization may fail if clock failure takes place or nasty misinformation from a group of nodes. The advantage is its computational complexity is low but for sensor devices energy efficiency is low.

### 3. *Timing-Sync Protocol for Sensor Network (TPSN)*

Synchronization based on sender–receiver approach is used. It supports both internal and external clock synchronization. Each node has a clock of 16-bit register and for identification unique ID is attached with each sensor. It performs synchronization in two steps: (1) Level Detection Phase (LDP) and (2) Synchronization Phase.

In level detection phase, the hierarchical topology of the network is generated. To build hierarchical topology for the network, a node at stage $S_i$ can connect with minimum one node at one stage below, i.e. at stage $S_{(i-1)}$. The algorithm is started by initiating broadcast message of level-detection phase. It is broadcasted by the root node. Child nodes are synchronized by the node at one level above in hierarchy. A spanning tree is constructed with a node which is selected as root. All pair of nodes can be considered as root and child, in which the child node may be referred as root for the next node. Child node initiates the request for synchronization to the root node. Root node replies with an acknowledgment message. Child node computes the clock offset relative to the root node based on arrival and departure time of each message [11].

In the synchronization step, TPSN carries out pair-wise synchronization with the hierarchical structure set up in the level-detection phase. Each node at stage $S_i$ synchronizes its clock with nodes at one stage below, i.e. stage $S_{(i-1)}$. It is shown in Fig. 1:

$t_1$: Node A is sender, starting sync by sending synchronization pulse packet to B.
$t_2 = t_1 + t_\Delta + d$, where $t_\Delta$ = clock offset and d = propagation delay.
$t_3$: B replies acknowledgment containing $t_1, t_2, t_3$.
$t_4$: A receives acknowledgement and $t_4 = t_3 - t_\Delta + d$.

**Fig. 1** TPSN message
transmission

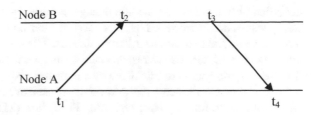

$$t_\Delta = [(t_2 - t_1) - (t_4 - t_3)]/2$$

$$d = [(t_2 - t_1) + (t_4 - t_3)]/2$$

After finding clock offset and delay in transmission, nodes use these values to correct its internal clock value. It achieves high energy efficiency and average computational complexity. There is a drawback that a conciliated node can cause an inaccurate offset value computed by its child node, hence tree will drop.

### 4. *Flooding Time Synchronization Protocol (FTSP)*:

FTSP is based on sender–receiver approach. It is implemented as an internal clock synchronization. It can also support external clock synchronization. At initial stage, root node broadcasts time synchronization message. Other nodes except root node coordinate root message with their neighbours. Non-root nodes use several neighbours' time stamp values to calculate their individual time. After a fixed time period, if there is no message for synchronization is received, a node can choose itself as root and repeat previous steps for synchronization [12]. The configuration of each node is as follows: Each node is assigned with a unique ID; each message intended for synchronization contains three fields: Time stamp, Root ID and Sequence Number. The node with the smallest ID will be only one root in the whole network. During the synchronization phase, root and synchronized node broadcast synchronization message. Nodes receive a synchronization message from the root or synchronized node. A node can estimate the offset by collecting sufficient synchronization messages. It calculates the offset and becomes a synchronized node by correcting its clock value. It achieves high energy efficiency and low computational complexity.

The drawback of FTSP is that if any compromised node can acquire the lowest ID, it always wins and hence can declare itself as a root. Due to the compromised node, other nodes may send incorrect data for time synchronization and hence fraudulent the entire tree.

### 5. *Time Diffusion Synchronization Protocol (TDSP)*:

TDSP is algorithm may implement synchronization based on internal synchronization approach. It is receiver-receiver synchronization algorithm. It is based on network-wide time synchronization protocol. TDSP works as follows [13]:

First step is Peer Evaluation Procedure (PEP) to assess the constancy of node's clock. It is assessed by using the Allan Variance [14]. It is also used to estimate the deviation of two clocks. At the initial stage, n numbers of messages with their time stamp values are broadcasted by the chosen master node. The neighbour nodes use it to calculate the two sample Allan Variances. These calculated variances are sent reverse to the master node as response message. Based on these values, clock deviations among the nodes are calculated by the master node. This procedure is repeated for n hopes of the master node. After PEP all sensor nodes receive average Allan Deviation. Next, follows Time Diffusion Procedure (TDP) in which the messages with timing information are diffused from the master nodes to neighbour nodes. These messages are more diffused by the selected diffused chief nodes. Effectively a hierarchical structure is produced. At the end, timing information is used to adjust the local clocks.

The advantage of TDSP is that it improves the performance of voice and video applications using directed diffusion in which multiple sensor nodes are sending data back to the sink. It can be applied for static and mobile sensor network. There is no need for precise time server. It achieves high energy efficiency and average computational complexity. One drawback is that the complexity of TDP becomes high as number of broadcast messages increases in each cycle. And another is if a clock value is accustomed to lesser clock value then time may run backward.

## 6. *Tiny-Sync and Mini-Sync*:

It is based on pair-wise synchronization with sender-to-receiver approach. It works based on internal clock synchronization and corrects local clock value. There is an assumption that the frequency of clocks in sensor networks is constant and has a linear correlation [7, 15]. Algorithms use two-way message interactions to accumulate data points and to estimate relative clock drift and clock offset. The relation between two cocks is defined as

$$T_1(t) = a_{12}T_2(t) + b_{12}$$

Where $a_{12}$ is a relative clock drift and $b_{12}$ is a relative clock offset. At time $t_0$, node $N_1$ transmits a time-stamped query and $N_2$ replies with time-stamped at time $t_1$. The arrival time of the responded message is recorded by $N_1$ at $t_2$. Which marks in data points $(t_0, t_1, t_2)$, which must satisfy $t_0 < a_{12}t_1 + b_{12}$ and $t_2 > a_{12}t_1 + b_{12}$. To increase the precision of algorithms, this procedure is reiterated several times to gain multiple data points and to find constraints on clock drift and clock offset value. All gathered data points are not useful.

Tiny-sync algorithm retains only four of these constraints. Each time when new record is available for data points then the existing four and the two new constraints are compared. From this record, only the best four results are kept. The disadvantage is those constraints that may give better measure but may be eliminated at the time of combining with other data points. In case of Mini-sync algorithm, it first makes sure that this data point is useless and then discards it. There is big computational cost and storage space cost, but precision is increased.

## 7. *PSync*:

PSync is the synchronization protocol based on visible light. It is based on receiver-to-receiver synchronization approach. It is implemented using internal clock synchronization but there is no clock correction is required. PSync relies on Light Emitting Diode (LED) for the light source and it is extremely energy efficient for the recipients. Idea is that pulsation of light formed by LED yield a proficient method for synchronizing close-by devices [16].

To synchronize devices, the minimal requirement for this protocol is to have a light sensor and need of a programmable source of light. Initially, to synchronize the message transmission time among the nodes, the preamble is used. An energy efficient preamble technique is used for synchronization [17]. This preamble should be small and priory it should inform the recipients about synchronization point. Within a small time period, recipients conclude the point of synchronization and again go for sleep until the point of synchronization. De Bruijn sequence is used for the preamble method [18]. Periodic sequences over a limited alphabet, in which each promising v-tuple appear just once in a given time, are known as De Bruijn sequences. It is used in position intellect applications. In this protocol, binary alphabet {0, 1} is used as a sequence. The synchronizing light sources produce a binary sequence of span v and with period of $2^v - 2$. The whole duration of preamble is $T_p = \tau * (2^v - 2)$. The recipient wakes up at regular intervals, at least one time whose length relies on the interval of preamble. A recipient accumulates s samples for all v symbols of the emitted binary sequence of a span v. A device translates its position in the preamble and then sleeps. For other preambles, recipient wakes up to identify the synchronization point precisely. At the point of synchronization, the devices sample at the most legitimated rate to reduce synchronization inaccuracy and achieves energy efficiency at high level. Computational complexity of this algorithm is also high.

## 4 Comparative Study of Synchronization Algorithms

Table 1 represents the comparison of time synchronization algorithms based on features and issues of time synchronization algorithms discussed. The algorithms are compared for evaluating performance based parameters such as computational complexity and energy efficiency [19, 20, 21].

## 5 Conclusion

The quality and performance of IoT-based applications are directly affected by time synchronization. For example for the sorting of events and to preserve the state of resources in smart logistic, smart banking, etc., time synchronization is necessary. This paper presents the study of algorithms for time synchronization which is applied for wireless sensor networks. The comparative study of various time synchronization algorithms will help in selecting time synchronization algorithms that can be applied

**Table 1** Analysis of time synchronization algorithms

| Name of algorithm | Performance parameters | | | | |
|---|---|---|---|---|---|
| | Internal versus external sync. | Approach | Clock correction | Energy efficiency | Computational complexity $n$ = No. of nodes |
| RBS | Both | Receiver-to-Receiver | Not required | High | $O(mn^2)$ m = No. of reference broadcast |
| LTS | Both | Sender-to-Receiver | Required | Low | $O(n^2)$ |
| FTSP | Both | Sender-to-Receiver | Required | High | $O(n)$ |
| TPSN | Both | Sender-to-Receiver | Required | High | $O(n)$ |
| TDSP | Internal | Receiver-to-Receiver | Required | High | $O(M + DL)$ M = No. of master nodes |
| Mini-sync | Internal | Sender-to-Receiver | Required | Low | It depends on no. of data points recorded. If m = no. of data points and N = no. of times message transmitted, then complexity is $O(mN)$ |
| Tiny-sync | Internal | Sender-to-Receiver | Required | Low | Same as Mini-Sync. $O(mN)$ |
| PSync | Internal | Receiver-to-Receiver | Not required | High | Time to generate De Bruijn sequence ($O(2^v)$, v = size of sequence) |

directly or after improvisation for IoT Applications or it may also become the basis for developing new and efficient algorithms for IoT applications. It is noted that the performance of discussed algorithms is directly affected by the energy efficiency of sensor nodes and time complexity of algorithms. In the case of IoT applications, there is a need to design a fast synchronization algorithm with high energy efficiency for better performance.

# References

1. Karen R., Scott E., Lyman C.: The internet of things: an overview. Internet Soc. (2015)
2. Recommendation ITU-T Y.2060: Overview of the internet of things, ITU-T telecommunication standardization of ITU (2012)
3. Fikret, S., Bulent, Y.: Time synchronization in sensor networks: a survey (2004)
4. Sommer, P., Wattenhofer, R.: Gradient clock synchronization in wireless sensor networks, In Proceedings of 8th ACM/IEEE International Conference on Information Processing in Sensor Networks (IPSN) (2009)
5. Jeremy, E., Lewis, G., Deborah, E.: Finc-grained network time synchronization using reference broadcasts. In: Proceedings of the Fifth Symposium on Operating systems Design and Implementation, Boston, MA (2002)
6. Terrell, RB., Nicholas, G., Roozbeh, J.: Data-driven synchronization for internet-of- things systems. ACM Trans. Embed. Comput. Syst. 16(3), Article 69 (2017)
7. Waltenegus, D., Christian, P.: Fundamentals of Wireless Sensor Network Theory and Practice. Wiley Publication, New York (2010)
8. Leandro, T.B., Edison, P.F., Tales, H.: Self-correcting time synchronization support for wireless sensor networks targeting applications on internet of things. Int. Federation Autom. Control. (hosted by Elsevier) (2016)
9. Cena, G., Scanzio, S., Valenzano, A., Zunino, C.: Implementation and evaluation of the reference broadcast infrastructure synchronization protocol. IEEE Trans. Ind. Inf. 11(3), 801–811 (2015)
10. Greunen, J.V., Rabaey, J.: Lightweight time synchronization for sensor networks. In: Proceedings of the 2nd ACM International Workshop on Wireless Sensor Networks and Applications (WSNA), San Diego, CA (2003)
11. Ganeriwal, S., Kumar, R., Srivastava, M.B.: Timing-sync protocol for sensor networks. In: Proceedings of 1st International Conference on Embedded Networked Sensor Systems (SenSys) (2003)
12. Miklós, M., Branislav, K., Gyula, S., Ákos, L.: The flooding time synchronization protocol. In: Proceedings of the 2nd ACM International Conference on Embedded Networked Sensor Systems (SenSys), Baltimore, MD, USA, pp. 39–49 (2004)
13. Weilian, S., Ian F.: Time diffusion synchronization protocol for wireless sensor networks. IEEE/ACM Trans. Netw. 13(2) (2005)
14. Allan, D.: Time and frequency (time-domain) characterization, estimation, and prediction of precision clocks and oscillators. IEEE Trans. Ultrason. Ferroelectr. Freq. Control. 34(6), 647–654 (1987)
15. Suyoung, Y., Chanchal, V., Mihail, S.: Tiny-sync: tight time synchronization for wireless sensor networks. ACM J. 5, 1–33 (2005)
16. XiangFa, G., Mobashir, M., Sudipta, S., Mun, C.C., Seth, G., Derek, L.: PSync: visible light based time synchronization for IoT. In: The 35th Annual IEEE International Conference on Computer and Communications (2016)
17. Romer, K.: Time Synchronization in ad hoc networks. In Proceedings of 2nd ACM International Symposium on Mobile Ad Hoc Networking & Computing (MobiHoc) (2001)

18. Mitchell, C.J., Etzion, T., Peterson, K.G.: A method for constructing decodable de Bruijn sequences. IEEE Trans. Inf. Theory **42**(5), 1472–1478 (1996)
19. Gopi, K.Y., Awadhesh, K., Akash, R.: Analysis of time synchronization protocols for wireless sensor networks: a survey. Int. J. Comput. Sci. Mob. Comput. **4**(5), 1062–1068 (2015)
20. Niranjan, P.: Consensus-based time synchronization algorithms for wireless sensor networks with topological optimization strategies for performance improvement (2016)
21. Aneeq, M., Thilo, S., Henning T., Reinhard, E.: Methods and performance aspects for wireless clock synchronization in IEEE 802.11 for the IoT. IEEE Trans. (2016)

# Comparative Study of the Electrical Energy Consumption and Cost for a Residential Building with Conventional Appliances Vis-a-Vis One with Energy-Efficient Appliances

Adeyemi Alabi, Oluwasikemi Ogunleye, Sanjay Misra,
Olusola Abayomi-Alli, Ravin Ahuja and Modupe Odusami

## 1 Introduction

This energy is an important recipe for the economic growth and development of nations. It is pertinent that energy is properly conserved in order to prevent avoidable waste of energy which will drastically reduce cost and the need to generate more energy to meet the national demand. Energy demand in Nigeria is expected to grow explosively in proportion to the anticipated rapid population growth between 2015 and 2050. The efficiency with which energy is used by firms and households has widespread impact on economic activity of the country, which in turn has implications for environmental quality and energy security. Buildings account for more than 40%

A. Alabi · O. Ogunleye · S. Misra (✉) · O. Abayomi-Alli · M. Odusami
Center of ICT/ICE Research, CUCRID Building, Covenant University, Ota, Nigeria
e-mail: sanjay.misra@covenantuniversity.edu.ng

A. Alabi
e-mail: adeyemi.alabi@stu.cu.edu.ng

O. Ogunleye
e-mail: oluwasikemi.ogunleye@stu.cu.edu.ng

O. Abayomi-Alli
e-mail: olusola.abayomi-alli@covenantuniversity.edu.ng

M. Odusami
e-mail: modupe.odusami@covenantuniversity.edu.ng

R. Ahuja
Department of Computer Engineering, University of Delhi,
New Delhi, India
e-mail: ravinahujadce@gmail.com

© Springer Nature Singapore Pte Ltd. 2019
R. K. Shukla et al. (eds.), *Data, Engineering and Applications*,
https://doi.org/10.1007/978-981-13-6351-1_24

297

of the global energy use today in both developed and developing countries [1]. Hence, it poses a major opportunity as a sector for consideration to reduce global energy consumption in Nigeria. A community with efficiently designed homes and offices will lower energy bills, liberate investible assets, and avoid unnecessary expenditure like building new power plants [2].

Energy efficiency in buildings has been relatively well researched, yet the problem of energy inefficient buildings in Africa still persists. The regulatory and voluntary approaches for enhancing building energy efficiency were reviewed in [3]. The opportunities and challenges to electrical energy conservation and $CO_2$ emissions reduction in Nigeria's building sector were evaluated in [4]. A paper review toward sustainable, energy-efficient, and healthy ventilation strategies in buildings was considered in [5].

Energy consumption in buildings includes lighting, domestic and commercial appliances, warming, ventilation, and cooling systems. Some inventive progresses have been made in energy-efficient buildings, and there is a progressing exertion in advancing and showing their energy savings' possibilities. However, the majority of these solutions are directed toward new buildings and future building designs, and is difficult to implement in existing buildings [5]. Although energy-efficient appliances promise savings in the cost incurred on energy consumptions, the replacement of old and inefficient appliances in old buildings in Nigeria is coming at a slow pace. This is because of the high cost that comes with replacing the appliances as most energy-efficient appliances are more costly than their equivalent inefficient appliances. It is therefore important to evaluate if the saving on energy consumptions by using an energy-efficient appliance will pay for the cost incurred to replace such appliance with an energy-efficient one within the expected life expectancy of the appliance. Against this background, this paper aims to design a framework to improve the energy efficiency of buildings in Nigeria to curb carbon emissions and increase energy savings. These innovative solutions will lead to sustainable development in the energy sector and the Nigerian economy. There is a lack of adequate literature to solve the inefficiency of buildings in the country. This paper therefore aims to design a framework for improving the energy efficiency of buildings in Nigeria.

This paper is structured as follows: The next section reveals some of the literature that were reviewed, followed by an explanation of the methodology implored and then a discussion of the result obtained and the paper was concluded.

## 2  Literature Review

According to [6], there is an inefficient utilization of available energy in Nigeria. Currently, most buildings in Nigeria lay more emphasis on the aesthetic values with little or no consideration for energy efficiency [7]. A more efficient utilization of energy resources can lessen greenhouse gas emissions and slow down depletion of nonrenewable energy resources [3].

Energy efficiency in buildings has been relatively well researched, yet the problem of energy inefficient buildings in Africa still persists. The regulatory and voluntary approaches for enhancing building energy efficiency were reviewed in [3]. They observed that the potential energy cost saving alone is an inadequate motivation to investing into improvement measures, unless there is an energy price shock. They recommended the adoption of a well-articulated policy mix involving both regulatory and voluntary instruments to achieve energy efficiency in buildings.

The opportunities and challenges to electrical energy conservation and $CO_2$ emissions reduction in Nigeria's building sector were evaluated in [4]. They found that putting all the energy saving opportunities they identified in place, at least 10% of total residential electrical energy use could be conserved while about 10% of both total industrial and commercial sectors electricity demand could be saved. These would significantly cut greenhouse gases emissions in the country. A framework of strategies to overcome these problems, encourage energy conservation, and thereby enhance sustainable development in Nigeria was then suggested. The need for energy-efficient buildings in Nigeria was evaluated in [8]. It was found that in Nigeria, most buildings do not take solar architecture and energy efficiency into consideration during construction due to ignorance, poverty, lack of awareness, and/or improper government policy on building regulations. The author addressed these issues by proffering solutions on the way forward for the country to achieve energy efficiency in buildings. A potential analysis of gray energy limits for residential buildings in Germany was performed in [9]. It was observed that the global warming potential (GWP) of shell constructions could be reduced by as much as 77% using existing technologies and with no additional investment costs. Environmental cost savings of more than €1 billion per year could be realized for investments in the German economy. With additional investments, the saving potential could jump to 95%.

A paper review toward sustainable, energy-efficient, and healthy ventilation strategies in buildings was considered in [5]. Evidence suggested that utilizing hybrid ventilation in buildings integrated with appropriate control strategies, to adjust between mechanical and natural ventilation, leads to substantial energy savings, while an appropriate indoor air quality is still maintained. A model-based optimization of distributed and renewable energy systems in buildings to address the design and control problem of building energy systems was developed in [10]. They developed a two-level optimization framework for the research. The results provided different optimal trade-off unit configurations with respect to the total investment cost and the defined self-sufficiency indicator. Despite the fact that this study solely considered typical Swiss residential dwellings, the presented framework could be applied in other types of buildings.

There is a lack of adequate literature to solve the inefficiency of buildings in the country. This paper therefore aims to design a framework for improving the energy efficiency of buildings in Nigeria.

# 3 Proposed Methodology

## 3.1 Case Study: Three-Bedroom Flat in Covenant University Ota

Covenant University is a growing community with over ten thousand people [11]. Domiciled in the mini-township is a gas-powered turbine. The energy use in the community ranges around 2 MW of peak load when the university is in full session and 1.25 MW of peak load when the university is not in session as shown in Fig. 1.

Figure 1 depicts a graphical representation comparing the peak loads in Covenant University when the school is on session and when out of session. Its graph shows that the peak load experienced during session is relatively high when compared to off session consumption. The electricity tariff for the residents of the Covenant University is ₦30 per kWh. Comparing this tariff with the current tariff set by the Nigerian Electricity Regulation Commission (NERC) effective from February 1, 2016, it can be seen that there is a tariff difference as shown in Table 1. This tariff in Table 1 is exclusively for residential houses.

Table 1 shows the energy cost of different classes of residence from 2015 to 2016 by Ikeja Distribution Company. The difference in the energy charges has increased tremendously from R2SP to R4 (H max demand—11/33 kV) class of residence ranging from 8 to 10.67 naira. The residents of this community enjoy 24-hour electricity supply and do not have to purchase or maintain backup generators like their counterparts living elsewhere, but it is also important for the members of this community to

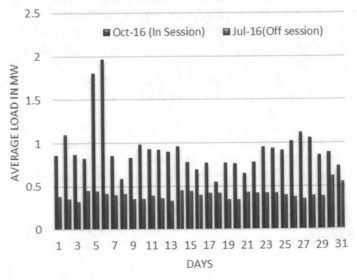

**Fig. 1** Graph showing a comparison of the peak loads experienced in Covenant University when the University is in session and out of session

**Table 1** Ikeja distribution company's energy charges in ₦/kWh [12]

| Class of residence | 2015 (Naira) | 2016 (Naira) | Naira difference in 2016 |
|---|---|---|---|
| R1 (Lifeline 50 kWh) | 4.00 | 4.00 | 0.00 |
| R2SP (Single phase) | 13.21 | 21.30 | 8.09 |
| R2TP (Three phase) | 13.21 | 21.80 | 8.59 |
| R3 (LV Maximum Demand) | 26.25 | 36.49 | 10.24 |
| R4 (HV Maximum Demand—11/33 kV) | 26.25 | 36.92 | 10.67 |

imbibe the energy efficiency paradigm. Not only will this reduce the cost of energy used but also it will make more energy available for the ever-increasing community.

## 3.2 Building Description (AS-IS)

- **Building envelope**: The building is 12-year-old, built with sandcrete blocks and has an aluminum roof finish.
- **HVAC system**: The building has a total of four-wall unit air conditioners located in various spaces of the house. All rooms have additional ceiling fans, while the living room has two ceiling fans.
- **Water management**: There is a centralized water pumping system for the members of the community. However, each of the houses has its own water heating system. In the audited house, the distributed water heating method was used and there is a total of three 15 L Ariston water heaters on in each of the bathrooms catering to the hot water needs of the family.
- **Appliances**: The appliances consuming the most electricity are the air conditioners and the water heaters. In most space in the building, incandescent bulbs were used for the lighting.

## 3.3 Energy Demand and Cost of the Building AS-IS

Based on the load schedule in Fig. 2 and the walk-through energy audit carried out on the residence, the energy demanded annually by each appliance installed was computed using Eqs. 2, 3 and using Eq. 4, and the energy cost was computed. Results can be found in Table 2. The following equations [13] are used to estimate the energy parameters:

## Duration of daily use (hours)

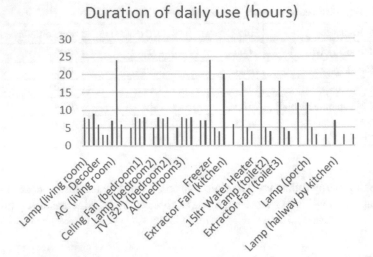

**Fig. 2** Load schedule of the residence

**Table 2** Comparison of the current energy consumption, retrofit energy consumption, and savings

| Model | Annual energy consumption (kWh) | Annual energy cost ( ₦ 30/kWh) |
|---|---|---|
| Building as-is | 80, 211.1225 | 2,406,333.68 |
| Building as-can-be | 42,052.38 | 1,261,571.4 |
| Savings | 38,158.7425 | 1,144,762.28 |

$$power = energy/time. \tag{1}$$

$$Energy\ demand\ (ED) = Power\ consumption\ x\ time\ taken \tag{2}$$

$$Annual\ Energy\ Demand\ (AED) = ED \times 365 \tag{3}$$

$$Annual\ Energy\ Cost\ (AEC) = AED \times tarrif/kWh \tag{4}$$

The installed energy demand in the residence (this is based on the appliances present in the house) is **80, 211.1225 KWhr** with an energy cost of **2,406,333.675 Naira**.

# 4 Comparison Result and Discussions

## 4.1 Recommendations of Energy-Efficient Methods (As-Can-Be)

Some of the proposed methods of energy efficiency for the audited building are as follows:

A. **Centralized water heating system**:

There are two major alternative solutions to the water heating consumption in the household:

- Centralized heating;
- More energy-efficient distributed water heaters.

Centralized heating is more energy-efficient as they provide hot water at the points of need. Also, the heat losses associated with storage tank distributed water heaters are also nonexistent. The centralized water heating was implored in our model. In place of the three distributed 15 L water heaters, one centralized 50 L water heater is to be used in our model to deliver hot water need to each of the three bathrooms.

B. **Appliances**: Replacing the appliances with energy star-rated devices will go a long way in increasing the energy efficiency of the household. The models of energy-rated refrigerators sold nowadays use less than half the amount of energy of models sold before 1993, effectively reducing the energy consumption by 50% [14]. Similarly, the air conditioners in the house can be retrofitted with more energy-efficient versions.

C. **Lighting**: Replacing the incandescent bulbs in the house with CFLs and LEDs should go a long way in reducing the energy consumption of the building by lighting. Homeowners have a proclivity of leaving lights in certain rooms on when not in use like the kitchen, store, and the restroom. Installing an occupancy in these rooms ensures that the lights are only on when they are occupied.

## 4.2 Energy Demand and Cost of the Retrofitted Model (As-Can-Be)

Based on the above-proposed solutions, an energy-efficient model of the building was designed. The inefficient appliances were substituted with energy-efficient appliances. The annual energy demand and cost was computed for the retrofitted model result as shown in Table 3; the estimated energy demand for the model building (as-can-be) is 42,052.38 KWh with the energy cost of 1,261,571.4 Naira. The retrofitted model in comparison to the building as is shown that by retrofitting the house with energy-efficient appliances 47.57% (38,158.7425 kWh) is saved yearly on energy demanded.

**Table 3** Table showing the annual energy demand and cost of each appliance in the building as-is and as-can-be

| S/n | Area | Appliance (type) | Appliance rat (W) current | Appliance rat (W) retrofit | Duration of use per day (h) | Annual energy demand (As-is) (kwh) | Annual energy demand (as-can-be) (kwh) | Annual energy savings (kwh) |
|---|---|---|---|---|---|---|---|---|
| 1 | Living room | Lamp | 60 × 10 | 20 × 26 | 8 | 1752 | 1518.4 | 233.6 |
| | | Celling fan | 75 × 2 | 30 × 2 | 7.5 | 410.63 | 164.25 | 246.375 |
| | | TV (42″) | 120 | 54 | 9 | 394.2 | 177.39 | 216.81 |
| | | Decoder | 100 | 100 | 6 | 219 | 219 | 0 |
| | | DVD | 50 | 50 | 3 | 54.75 | 54.75 | 0 |
| | | Sound system | 150 | 65 | 3 | 164.25 | 71.175 | 93.075 |
| | | AC | 2700 | 1340 | 7 | 6898.5 | 3423.7 | 3474.8 |
| | | Dispenser | 580 | 580 | 24 | 5080.8 | 5080.8 | 0 |
| | | Electric iron | 1500 | 1500 | 6 | 3285 | 3285 | 0 |
| | | Lamp | 40 × 4 | 20 × 6 | 5 | 292 | 219 | 73 |
| 2 | BedRoom1 | AC | 2700 | 1340 | 8 | 7884 | 3912.8 | 3971.2 |
| | | Celling fan | 75 | 30 | 7.5 | 205.31 | 82.125 | 123.1875 |
| | | TV (32″) | 105 | 31 | 8 | 306.6 | 90.52 | 216.08 |
| | | Lamp | 40 × 4 | 20 × 6 | 5 | 292 | 219 | 73 |
| 3 | BedRoom2 | AC | 2700 | 1340 | 8 | 7884 | 3912.8 | 3971.2 |
| | | Celling fan | 75 | 30 | 7.5 | 205.31 | 82.125 | 123.1875 |
| | | TV (32″) | 105 | 31 | 8 | 306.6 | 90.52 | 216.08 |
| | | Lamp | 40 × 4 | 20 × 6 | 5 | 292 | 219 | 73 |

(continued)

**Table 3** (continued)

| S/n | Area | Appliance (type) | Appliance rat (W) current | Appliance rat (W) retrofit | Duration of use per day (h) | Annual energy demand (As-is) (kwh) | Annual energy demand (as-can-be) (kwh) | Annual energy savings (kwh) |
|---|---|---|---|---|---|---|---|---|
| 4 | BedRoom3 | AC | 2700 | 1340 | 8 | 7884 | 3912.8 | 3971.2 |
| | | Celling fan | 75 | 30 | 7.5 | 205.31 | 82.125 | 123.1875 |
| | | TV (32″) | 105 | 31 | 8 | 306.6 | 90.52 | 216.08 |
| | | Lamp | 40 × 2 | 20 × 3 | 7 | 204.4 | 153.3 | 51.1 |
| | | AC | 2700 | 1340 | 7 | 6898.5 | 3423.7 | 3474.8 |
| | | Freezer | 1600 | 500 | 24 | 14016 | 4380 | 9636 |
| 5 | Kitchen | Elect. kettle | 2000 | 2000 | 5 | 3650 | 3650 | 0 |
| | | Microwave | 1400 | 1000 | 4 | 2044 | 1460 | 584 |
| | | Ext. fan | 6 | 6 | 20 | 43.8 | 43.8 | 0 |
| 6 | Store | Lamp | 20 | 8 | 6 | 43.8 | 17.52 | 26.28 |
| | | Ext. fan | 6 | 6 | 18 | 39.42 | 39.42 | 0 |
| 7 | Toilet1 | Lamp | 40 | 20 | 5 | 73 | 36.5 | 36.5 |
| | | H$_2$O heater | 1500 | 0 | 4 | 2190 | 0 | 2190 |
| | | Ext. fan | 6 | 6 | 18 | 39.42 | 39.42 | 0 |
| 8 | Toilet2 | Lamp | 40 | 20 | 5 | 73 | 36.5 | 36.5 |
| | | H$_2$O heater | 1500 | 0 | 4 | 2190 | 0 | 2190 |
| | | Ext. fan | 6 | 6 | 18 | 39.42 | 39.42 | 0 |

(continued)

**Table 3** (continued)

| S/n | Area | Appliance (type) | Appliance rat (W) current | Appliance rat (W) retrofit | Duration of use per day (h) | Annual energy demand (As-is) (kwh) | Annual energy demand (as-can-be) (kwh) | Annual energy savings (kwh) |
|---|---|---|---|---|---|---|---|---|
| 9 | Toilet3 | Lamp | 40 | 20 | 5 | 73 | 36.5 | 36.5 |
| | | H$_2$o heater | 1500 | 0 | 4 | 2190 | 0 | 2190 |
| 10 | Terrace | Lamp | 40 × 2 | 20 × 3 | 12 | 350.4 | 262.8 | 87.6 |
| | | Lamp | 40 | 20 | 12 | 175.2 | 87.6 | 87.6 |
| 11 | Porch | H$_2$O pump | 400 | 400 | 5 | 730 | 730 | 0 |
| | | Washing machine | 500 | 500 | 3 | 547.5 | 547.5 | 0 |
| 12 | Ent. Porch | Lamp | 40 × 2 | 20 × 3 | 3 | 87.6 | 65.7 | 21.9 |
| 13 | Hallway1 | Lamp | 40 | 20 | 7 | 102.2 | 51.1 | 51.1 |
| 14 | Hallway2 | Lamp | 40 | 20 | 3 | 43.8 | 21.9 | 21.9 |
| 15 | Hallway3 | Lamp | 40 | 20 | 3 | 43.8 | 21.9 | 21.9 |
| Total energy (kWh) | | | | | | 80, 211.1225 | 42,052.38 | 38,158.7425 |
| Total cost (Naira) | | | | | | 2,406,333.68 | 1,261,571.4 | 1,144,762.28 |

## 4.3 Comparative Analysis of the Current Model and the Energy-Efficient Model

The total power demand from the current energy consumption model is about 80, 211.12 kWh, while the energy demand using the energy-efficient method (Building as-can-be) is 42,052.38 kWh. This results in a considerable decrease in the energy demanded and used by the household, and about 38,158.7425 kWh of energy saved yearly. This amount of energy saved can be used elsewhere, thereby reducing the need to create more power plants and eventually reducing carbon emission in the country. The total cost of running the current energy consumption model amounts to about ₦2,406,333.68 spent yearly. By implementing the retrofit/energy-efficient method, this could be reduced to ₦1,261,571.4 such that ₦1,144,762.28 would be saved per year Fig. 3.

## 4.4 Payback Period (PBP) of Retrofit Appliances

Energy-efficient appliances usually cost more than non-energy-efficient appliances. Before retrofitting is performed in a building, it is important to investigate and compare the payback period of the appliance against the life expectancy of that appliance. Retrofitting is usually performed when the payback period of the appliance is lesser than its life expectancy [15, 16]. The payback period of an item is given as the ratio between cost of purchasing the unit and the cost of energy saved per year. A breakdown of the PBP of some targeted electrical appliances in the house is given in Table 4. It can be seen that all of the retrofitted items provide more energy savings in the long run. Items like the energy-rated star ceiling fan in Bedroom1, which may cost about ₦9800, could save as much as 123.19 kWh/year and ₦3695.63 in energy

**Fig. 3** Comparison between energy consumption and energy savings

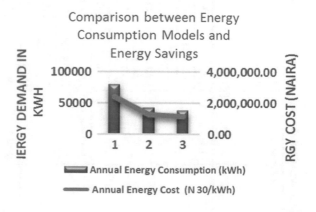

**Table 4** Retrofitted electrical appliances, energy demand savings per year, energy cost savings per year

| S/n | Area | Appliance (type) | Building (As-is) appliance rat (W) | Building appliance rat (W) | Cost of annual energy saved | Cost of purchasing energy-efficient appliance | Payback period (years) | Life expectancy (years) | Decision |
|---|---|---|---|---|---|---|---|---|---|
| 1 | Living room | Lamp | 60 × 10 | 20 × 26 | 7,008.00 | 21,021.00 | 3.00 | 2.74 | FALSE |
| | | Ceiling fan | 75 × 2 | 30 × 2 | 7,391.25 | 9,800.00 | 1.33 | 13.00 | TRUE |
| | | TV (42″) | 120 | 54 | 6,504.30 | 135,000.00 | 20.76 | 18.00 | FALSE |
| | | Sound system | 150 | 65 | 2,792.25 | 42,800.00 | 15.33 | 15.00 | FALSE |
| | | AC | 2700 | 1340 | 104,244.00 | 108,500.00 | 1.04 | 13.00 | TRUE |
| | | Lamp | 40 × 4 | 20 × 6 | 2,190.00 | 4,851.00 | 2.22 | 4.38 | TRUE |
| 2 | BedRoom1 | AC | 2700 | 1340 | 119,136.00 | 108,500.00 | 0.91 | 13.00 | TRUE |
| | | Ceiling fan | 75 | 30 | 3,695.63 | 9,800.00 | 2.65 | 13.00 | TRUE |
| | | TV (32″) | 105 | 31 | 6,482.40 | 80,000.00 | 12.34 | 18.00 | TRUE |
| | | Lamp | 40 × 4 | 20 × 6 | 2,190.00 | 4,851.00 | 2.22 | 4.38 | TRUE |
| 3 | BedRoom2 | AC | 2700 | 1340 | 119,136.00 | 108,500.00 | 0.91 | 13.00 | TRUE |
| | | Ceiling fan | 75 | 30 | 3,695.63 | 9,800.00 | 2.65 | 13.00 | TRUE |
| | | TV (32″) | 105 | 31 | 6,482.40 | 80,000.00 | 12.34 | 18.00 | TRUE |
| | | Lamp | 40 × 4 | 20 × 6 | 2,190.00 | 4,851.00 | 2.22 | 4.38 | TRUE |
| 4 | BedRoom3 | AC | 2700 | 1340 | 119,136.00 | 108,500.00 | 0.91 | 13.00 | TRUE |
| | | Ceiling fan | 75 | 30 | 3,695.63 | 9,800.00 | 2.65 | 13.00 | TRUE |

(continued)

**Table 4** (continued)

| S/n | Area | Appliance (type) | Building (As-is) appliance rat (W) | Building appliance rat (W) | Cost of annual energy saved | Cost of purchasing energy-efficient appliance | Payback period (years) | Life expectancy (years) | Decision |
|---|---|---|---|---|---|---|---|---|---|
| | | TV (32″) | 105 | 31 | 6,482.40 | 70,000.00 | 10.80 | 18.00 | TRUE |
| | | Lamp | 40 × 2 | 20 × 3 | 1,533.00 | 2,425.50 | 1.58 | 3.13 | TRUE |
| | | AC | 2700 | 1340 | 104,244.00 | 108,500.00 | 1.04 | 13.00 | TRUE |
| 5 | Kitchen | Freezer | 1600 | 500 | 289,080.00 | 115,200.00 | 0.40 | 15.00 | TRUE |
| | | Microwave | 1400 | 1000 | 17,520.00 | 58,000.00 | 3.31 | 10.00 | TRUE |
| 6 | Store | Lamp | 20 | 8 | 788.40 | 1,960.00 | 2.49 | 11.42 | TRUE |
| 7 | Toilet1 | Lamp | 40 | 20 | 1,095.00 | 808.50 | 0.74 | 4.38 | TRUE |
| 8 | Toilet2 | Lamp | 40 | 20 | 1,095.00 | 808.50 | 0.74 | 4.38 | TRUE |
| 9 | Toilet3 | Lamp | 40 | 20 | 1,095.00 | 808.50 | 0.74 | 4.38 | TRUE |
| 10 | Terrace | Lamp | 40 × 2 | 20 × 3 | 2,628.00 | 2,425.50 | 0.92 | 1.83 | TRUE |
| 11 | Porch | Lamp | 40 | 20 | 2,628.00 | 808.50 | 0.31 | 1.83 | TRUE |
| 12 | Entrance Porch | Lamp | 40 × 2 | 20 × 3 | 657.00 | 2,425.50 | 3.69 | 7.31 | TRUE |
| 13 | Hallway1 | Lamp | 40 | 20 | 1,533.00 | 808.50 | 0.53 | 3.13 | TRUE |
| 14 | Hallway2 | Lamp | 40 | 20 | 657.00 | 808.50 | 1.23 | 7.31 | TRUE |
| 15 | Hallway3 | Lamp | 40 | 20 | 657.00 | 808.50 | 1.23 | 7.31 | TRUE |

cost per year. The life expectancy of the ceiling fan is 13 years, and the expected payback period calculated with Eq. 4 for the ceiling fan is 2.65 years. Since the life expectancy of the appliance is greater than its payback period, it is economically viable to substitute the current appliance with an energy-rated one. This means that in about two and half years, the ceiling fan would have paid for itself through its savings on energy cost. After which it will continue to save energy cost. The life expectancy of an item is subject to how to appliance is used and maintained by the owners. If an appliance is used and maintained properly, it may last longer than its initial design considerations [17]

$$PBP = \text{Cost of purchasing unit/Energy cost saved per year} \qquad (5)$$

The payback period of each substituted appliance was calculated using Eq. 5. An interactive MATLAB program (Appendix A) was written to help users determine whether or not it is economically viable to buy any energy-efficient appliance.

After calculating the payback period for each energy-efficient appliance, the payback period in years was compared to the life expectancy in years given by the manufacturer to see if it is economically viable to purchase the appliance or not. It was determined that when the life expectancy is greater than the payback period, it is economically viable to purchase the appliance; otherwise, it is not. The decision column in Table 4 shows which appliance is economically viable to purchase, where "TRUE" means to purchase and "FALSE" means do not purchase.

# 5 Conclusion

This paper reveals more insight into how an as of now existing building can be retrofitted to be more energy-efficient. A stroll through energy review was done taking a standard three-bedroom flat in Covenant University as contextual analysis. It is trusted that by adopting energy-efficient technologies and retrofitting the house in that order, family units in Covenant University will have the capacity to spare more cash over the long haul, and furthermore this will come about into more energy for use in the community. The retrofitted model demonstrates that about 47.5% of the cost of energy utilized yearly can be saved. By replicating this model to every one of the family units inside the community, more energy would be spared all things considered and there would be little need to expand the generating limit of the community later on even with the energy requirements of its regularly expanding populace.

# References

1. Ashrafian, T., Yilmaz, A.Z., Corgnati, S.P., Moazzen, N.: Methodology to define cost-optimal level of architectural measures for energy efficient retrofits of existing detached residential buildings in Turkey. Energy Build. Elsevier. **120**, 58–77 (2016)
2. Hafemeister, A..R.a.D.: Energy-efficient Buildings. In Energy in Buildings, California, CaliTech, **2**, pp. 1–8 (1988)
3. Lee, W.L., Yik, F.W.H.: Regulatory and voluntary approaches for enhancing building energy efficiency. Prog. Energy Combust. Sci. **30**, 477–499 (2004)
4. Akinbami, J.F., Lawal, A.: Opportunities and challenges to electrical energy conservation and $CO_2$ emissions reduction in Nigeria's building sector, cities and climate change, vol. 2, pp. 345–365 (2009)
5. Chenari, B., Carrilho, J.D., da Silva, M.G.: Towards sustainable, energy-efficient and healthy ventilation strategies in buildings: a review. Renew. Sustain. Energy Rev. **59**, 1426–1447 (2016)
6. Oyedepo, O.: Efficient energy utilization as a tool for sustainable development in Nigeria. Int. J. Energy Environ. Eng. **3**(11), 1–12 (2012)
7. Oyedepo, S.: On energy for sustainable development in Nigeria. Renew. Sustain. Energy Rev. **16**, 2583–2598 (2012)
8. Nwofe, P.: Need for energy efficient buildings in Nigeria. Int. J. Energy Environ. Res. **2**(4), 1–9 (2014)
9. Brinks, P.: Potential-analysis of grey energy limits for residential buildings in Germany. Energy Build. **127**, 580–589 (2016)
10. Stadler, P., Ashouri, A., Maréchal, F.: Model-based optimization of distributed and renewable energy systems in buildings. Energy Build. **120**, 103–113 (2016)
11. Atayero, A.: SmartCU (Ota Local Government) (2 Aug 2016). https://www.us-ignite.org/globalcityteams/actioncluster/cKdkmgRELh3fn4RXrnyVzi/
12. Nigerian Electricity Regulatory Commission, "Multi-year Tariff Order (MYTO)—2015 for Ikeja Electricity Distribution Company (IKEDC) for the period 1st January 2015 to Decemeber 2024," Nigerian Electricity Regulatory Commission. Retrived from: http://www.nercng.org/index.php/media-and-publicity/public-notices/326-notice-of-commencement-of-myto-2015, Abuja, 2015
13. Wara, S.T., Abayomi-AIIi, A., Mohamed, A.K., Kahn, M.T.E.: Investigating electricity cost savings in Igbinedion University Campuses. In: 16th International Conference on Industrial and Commercial Use of Energy, Cape Town, South Africa (2008)
14. Krarti, M.: Energy Audit of Building Systems: An Engineering Approach. CRC Press Taylor & Fancis Group, United States of America (2011)
15. Robert Thornton, D.B.M.M.D.: Minimum household Energy Need, Vincentian Partnership for Social Justice (2014)
16. Rosenquist, G., Coughlin, K., Dale, L., McMahon, J., Meyers, S.: Life-cycle cost and payback period analysis for commercail unitary air conditioners. Ernest Orlando Lawrence Berkeley National Laboratory, California (2004)
17. National Association of Home Builders/Bank of America Home Equity, Study of Life Expectancy of Home Components, National Association of Home Builders, U.S.A (2007)

# Particle Swarm Optimization-Based MPPT Controller for Wind Turbine Systems

**Shefali Jagwani and L. Venkatesha**

## 1 Introduction

The maximum power which can be extracted from a wind turbine mainly depends on two factors, viz., velocity of wind and operating point of the system. Therefore, MPPT is an important step to increase the efficiency of the system [1]. The study of these MPPT techniques shows that they may actually fall in five categories, namely optimal torque control (OP), hill climbing search control (HCS), power signal feedback control (PSF), tip-speed ratio control (TSR), and power mapping control [1, 2]. Many papers have been published to study these techniques with their advantages and disadvantages [1–3]. In this paper, TSR control method is implemented (Fig. 1) using particle swarm optimization (PSO), which reduces the oscillations around the optimal point and increases the efficiency of wind turbine [4, 5]. This artificial intelligence technique is found to be more efficient and faster than other techniques.

One of the important requirements of TSR control is an anemometer which measures the wind velocity. The preknown values of the optimal TSR are stored in lookup table. Then, the measured wind velocity is converted to its corresponding optimal speed reference with the help of these values [6]. This method offers faster control and it is possible to yield more energy. PSO-based TSR control method helps to eliminate the requirement of any prior information about the system and calculates the optimal value online, and thus helps in extracting the peak power from the turbine.

S. Jagwani (✉) · L. Venkatesha
BMSCE, Bangalore, India
e-mail: shefalijagwani@gmail.com

L. Venkatesha
e-mail: lvenkatesha.eee@bmsce.ac.in

© Springer Nature Singapore Pte Ltd. 2019
R. K. Shukla et al. (eds.), *Data, Engineering and Applications*,
https://doi.org/10.1007/978-981-13-6351-1_25

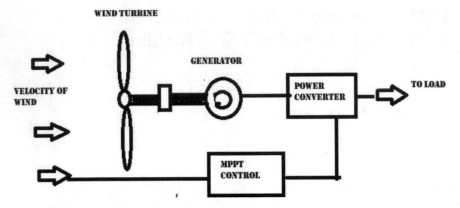

**Fig. 1** MPPT control for wind turbine system

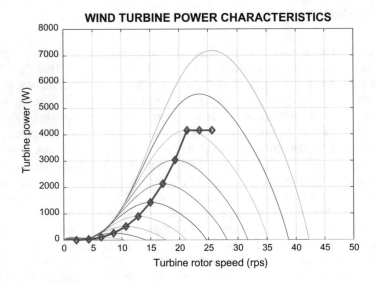

**Fig. 2** Wind turbine characteristics

## 2 Wind Turbine Characteristics

The mechanical power extracted from the wind generators $P_m$ (as shown in Fig. 2) is expressed as [1, 2]

$$P_m = \frac{1}{2}\rho A v^3 C_p(\lambda, \beta) \tag{1}$$

**Fig. 3** Power coefficient
curve

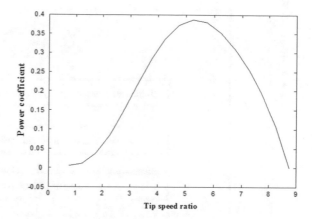

where

| | |
|---|---|
| $\rho$ | air density (kg/m$^3$), |
| $v$ | wind speed (m/s), |
| A | area swept by the rotor blades (m$^2$) |
| $C_p(\lambda, \beta)$ | coefficient of performance |

The parameter $C_p$ defines the power extraction efficiency. It is a nonlinear function of TSR ($\lambda$) and pitch angle of the blade ($\beta$). In this work, a variable speed fixed pitch turbine is considered, hence $\beta$ is fixed. The maximum theoretical value of performance coefficient $C_p$ is 0.59 according to the Betz limit. TSR is the ratio of rotational speed of turbine ($\omega$) to linear speed of blade tips and is given by [7]

$$\lambda = \omega R / v \tag{2}$$

where R = radius of blade. Different versions of the equation for obtaining $C_p$ have been defined by the authors in previously published papers [1–3, 8, 9]. In this paper, the following equation is used [10]:

$$C_p = 0.00234 + 0.03227\lambda - 0.076354\lambda^2 + 0.061857\lambda^3$$
$$- 0.015155\lambda^4 + 0.001514\lambda^5 - 0.000055\lambda^6 \tag{3}$$

Figure 3 shows the simulated plot of power coefficient $C_p$ with TSR. It can be observed that the maximum $C_p$ (0.392) occurs at $\lambda = 5.278$ and $\beta = 0°$. The torque of the turbine is stated as

$$T_t = P_t / \omega_t \tag{4}$$

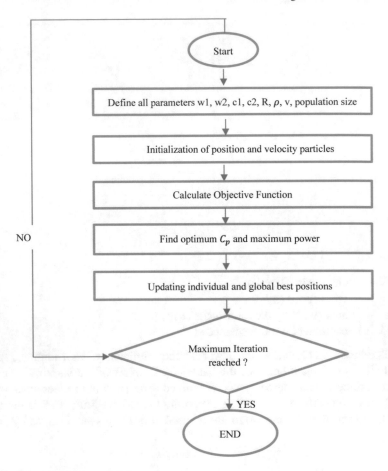

**Fig. 4** Flowchart for PSO algorithm

## 3   Particle Swarm Optimization

PSO technique is a population-based algorithm which was first presented by Kennedy and Eberhart in 1995. This technique is influenced by the study of artificial livings [11]. It can attain high-quality solutions with shorter time when compared to other heuristic methods, and therefore is used for many engineering and control systems [12–15]. The swarm-based process for searching is analogous to birds migrating in a group toward one destination, where the intelligence and cooperation of the whole flock result in efficient system. It is essential to note that they should be able to move in synchronism without colliding. The system is first initialized with randomly generated solutions known as particles, and then performs the iterative search for the optimum solution of the problem. The procedure is explained [16] in the flowchart (Fig. 4). The velocity of the particle is updated using Eq. (5) and Eq. (6), respectively:

**Table 1** Parameters used in PSO program

| Parameters | Value |
|---|---|
| Population size | 30 |
| Dimension | 1 |
| Maximum iterations | 30 |
| Wmax | 9 |
| Wmin | 0.4 |
| C1 | 2.0 |
| C2 | 2.0 |
| $\rho$ | 1.08 |
| $\beta$ | 0 |
| R | 2.5 |

$$v_{k+1}^i = w v_k^i + C_1 r_1 \left( pbest^i - x_k^i \right) + C_2 r_2 \left( gbest_k - x_k^i \right) \tag{5}$$

$$x_{k+1}^i = x_k^i + V_{k+1}^i \tag{6}$$

where $v_{k+1}^i$ is updated velocity, $w$ is momentum factor, $v_k^i$ is current velocity, $C_1$ and $C_2$ are acceleration constants, $r_1, r_2$ are the random values from 0 to 1, $pbest^i$ is the individual best value, $gbest_k$ is the global best value, $x_k^i$ is current position, and $x_{k+1}^i$ is modified position.

The objective function is defined using Eq. (3) and with the optimal value of $C_p$, maximum power for each wind velocity can be found out. The parameters which are used in this proposed technique are shown in Table 1.

## 4 PSO Simulation Results

For tracking the maximum value of power, the controller should adjust the velocity of wind generator according to the available wind velocity. The proposed MPPT technique uses wind power coefficient $C_p$ as the objective function. There is an optimum TSR value that makes power coefficient, $C_p$ maximum and hence, the maximum power is captured. Therefore, the key factor to obtain the maximum power is to optimize the $C_p$ by optimizing TSR. The simulation was performed for various wind speeds ranging from 8 m/s to 14 m/s. The value of $C_p$ for all wind speeds is shown in Table 2. From the results, it is observed that the PSO algorithm rapidly tracks the $C_p$ value, for which the output power is maximum. The error can be reduced by increasing number of iterations and time duration of the simulation.

**Table 2** Optimization results for wind power coefficient

| S. No. | Wind speed (m/s) | Optimal Wind power coefficient using PSO ($C_p$) | Wind power coefficient from equation |
|---|---|---|---|
| 1. | 8 | 0.3920 | 0.392 |
| 2. | 9 | 0.3921 | 0.392 |
| 3. | 10 | 0.3921 | 0.392 |
| 4. | 11 | 0.3910 | 0.392 |
| 5. | 12 | 0.3905 | 0.392 |
| 6. | 13 | 0.3670 | 0.392 |
| 7. | 14 | 0.3530 | 0.392 |

## 5 Conclusion

An intelligent control algorithm for tracking the maximum power in a wind turbine is proposed in this paper, which is an important requirement for obtaining high efficiency. This method is advantageous over other methods as it is simple and adaptive. It tracks the power with minimum oscillations at the peak value. Hence, one can conclude that if TSR is always preserved at the optimal value, the maximum energy can be extracted. This proposed algorithm is capable of deciding the optimum operating point with maximum power. The simulation results verify that this proposed technique provides an efficient MPPT-based controller for wind turbine and helps in efficient renewable energy generation.

## References

1. Kazmi, S.M.R., Goto, H., Guo, H.-J., Ichinokura, O.: Review and critical analysis of the research papers published till date on maximum power point tacking in wind energy conversion system. In: 2010 IEEE Energy Conversion Congress and Exposition (ECCE), pp. 4075–4082 (2010)
2. Abdullah, M.A., Yatim, A.H.M., Tan, C.W.: A study of maximum power point tracking algorithms for wind energy system. In: 2011 IEEE First Conference on Clean Energy and Technology, Kuala Lumpur, Malaysia, pp. 321–326 (2011)
3. Hui, J., Bakhshai, A.: A new adaptive control algorithm for maximum power point tracking for wind energy conversion systems. In: IEEE Power Electronics Specialists Conference, Greece, pp. 4003–4007 (2008)
4. Naveen Ram, G., Kiruthiga, A., Devi Shree, J.: A novel maximum power point tracking system for wind energy conversion system using particle swarm optimization. Int. J. Eng. Res. Technol. **3**(2), 1577–1581 (2014)
5. Santhana Krishnan, T., Sharmeela, C.: PSO based MPPT controller for wind energy system with high gain resonant converter feeding micro grid. Int. J. Latest Trends Eng. Technol. **6**(2), 181–191 (2015)

6. Maheswari, K., Porselvi, T.: Comparison of TSR and PSO based MPPT algorithm for wind energy conversion system. Int. J. Adv. Res. Electr., Electron. Instrum. Eng. **3**(Special Issue 1), 160–163 (2014)
7. Das, K.K., Buragohain, M.: An algorithmic approach for maximum power point tracking of wind turbine using particle swarm optimization. Int. J. Adv. Res. Electr., Electron. Instrum. Eng. **4**(5), 4100–4106 (2015)
8. Sarvi, M., Abdi, S., Ahmadi, S.: A new method for maximum power point tracking of PMSG wind generator using PSO fuzzy logic. Tech. J. Eng. Appl. Sci. **3**, 1984–1995 (2013)
9. Abdullah, M.A., Yatim, A.H.M., Tan, C.W., Samosir, A.S.: Particle swarm optimization—based maximum power point tracking algorithm for wind energy conversion system. In: 2012 IEEE International Conference on Power and Energy (2012)
10. Raju, A.B., Chatterjee, K., Fernandes, B.G.: A simple maximum power point tracker for grid connected variable speed wind energy conversion system with reduced switch count power converter. In: 2003 IEEE 34th Annual Power Electronics Specialist Conference (PESC—03), vol. 2 (2003)
11. Kennedy, J., Eberhart, R.: Particle swarm optimization. In: Proceedings, IEEE Conference on Neural Networks, 1995, vol. 4, pp. 1942–1948 (1995)
12. Evangeline, J., Cynthia, J., Darwin, J.D., Jeyanthy, P.A., Devika, S.: Power coefficient in wind power using particle swarm optimization. In: International Conference on Control, Instrumentation, Communication and Computational Technologies, pp. 71–75 (2014)
13. Kesraoui, M., Korichi, N., Belkadi, A.: Maximum power point tracker of wind energy conversion system. J. Renew. Energy, Elsevier **36**, 2655–2662 (2011)
14. Rathi, R., Sandhu, K.S.: Comparative analysis of MPPT algorithms using wind turbines with different dimensions & ratings. In: 1st IEEE International Conference on Power Electronics. Intelligent Control and Energy Systems (2016)
15. Sabazevri, S., Karimpour, A., Monfared, M., Bagher M., Sistani, N.: MPPT control of wind turbines by direct adaptive fuzzy-PI controller and using ANN-PSO wind speed estimator. J. Renew. Sustain. Energy **9**(1) (2017)
16. Blondin, J.: Particle swarm optimization: a tutorial. http://www.cs.armstrong.edu/saad/csci8100/pso_tutorial.pdf

# A New Model of M-secure Image Via Quantization

Vijay Bhandari, Sitendra Tamrakar, Piyush Shukla and Arpana Bhandari

## 1 Introduction

Steganography [1–4] and steganalysis [5] have more focus in the fields concerning law enforcement [6–8] and national strategic defence. It is carving and mastery [9] of the uncover secret intelligence in the cover forum [10, 11]. Contrary between steganalysis and steganography [12, 13] is the exposure of masked secret data embedded in the cover media also called as stego image. Image encryption [9, 10] makes use of the natural possessions of an image [1, 6], such as high dismissal and well-built spatial correlation. The encryption technique protects illegal access of the data. The encrypted image [9, 14] is a noisy image such that no one can obtain the secret image [2, 9] data without the correct key. The steganography [8] contains hidden a digital picture into another cover multimedia data such as image [9] and video. Steganography technique [1–3, 6] is used when encryption [3, 4, 6] is not acceptable. The purpose of steganography [8] embeds secret data in the reselected image [9]. The structure of the paper consists of mosaic figure generation and existing work, proposed work, and the results are compared based on different parameters and conclusion.

V. Bhandari (✉)
AISECT University Bhopal, Raisen, India
e-mail: vphd2k14@gmail.com

S. Tamrakar
Christian College, Chennai, India
e-mail: drsitendra@gmail.com

P. Shukla
UIT RGPV, Bhopal, India
e-mail: pphdwss@gmail.com

A. Bhandari
CSE Department, SIRTE, Sirte, Libya
e-mail: arpanabhandari08@gmail.com

© Springer Nature Singapore Pte Ltd. 2019
R. K. Shukla et al. (eds.), *Data, Engineering and Applications*,
https://doi.org/10.1007/978-981-13-6351-1_26

## 1.1 Mosaic Figure Production and Uncover Picture Recovery Information

### 1.1.1 Opting Suitable End Blocks for Every Pantile

Mosaic expression is the process of producing a picture or other florid items by combining together tiny chunks of stones and glass. The term mosaic keeps a number of tiny images called "Tiles" and is then placed on the single image called "TARGET image", before creating a mosaic image, there must be tiles embedded into the target image. The embedding of the tiles should be in the format [13, 15].

### 1.1.2 Mosaic Creation Using Different Similarity Measures

The source image is said to be secretly embedded in the resulting mosaic image. For creating a mosaic, the target image and the secret image can be split, but the size of splitting is different. The secret image is split for merging into a target image; the target image is split for finding the absolute place to fit the tile image in the target image [11]. In detail, the target image is split into very small tiny images and based on this, each tile image is compared with every target split image, and finds the best fit for that tile image, and in that position, the tile image is placed and so on. Generated mosaic image is the same as the target image with a little bit of quality drop. Mosaic image quality is high when splitting of the source image is high. The major difficulty is retrieving the tile images from the created mosaic images. This required the specific efficient technique for getting the secret image from the mosaic image. The creation process is done in a particular place and the retrieving process is done at another place, and for this, they can give the detail of which tile is stored in which position and the name of the image. This information is also sent to the other user so they can also be able to retrieve the image from the mosaic image [9] (Fig. 1).

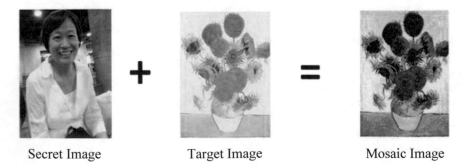

Secret Image          Target Image          Mosaic Image

**Fig. 1** Mosaic creation using different similarity measures

### 1.1.3 Background of Secret Image Recovery

The embedded information we have to extract to recover nearly lossless the secret image from the generated mosaic image [4].

## 2 Existing Work

This paper [4] introduced a colored black-and-white visual cryptography scheme, which adopts colored pixels [4] in dark descriptions to share a black and sallow top secret image. This paper [3] represents an algorithm to reversibly place a clandestine information in encrypted similes is presented, which consists of picture encryption, information embedding, and data taking out [3].

The new secret image [2] sharing scheme identifying the continuation of cheater is introduced and analyzed in this paper. A method to ensure the honesty of clandestine picture prior to its improvement is proposed.

This paper represents proposed here [6], which transforms robotically a given large-volume secret depiction into a so-called secret-fragment-visible [9, 10, 14] mosaic image of the same size [10, 14]. The mosaic image, which looks similar to an randomly elected end image and may be used as a disguise of the secret image, is yielded by dividing the clandestine image into wreckage and transforming their color uniqueness to be those of the corresponding blocks of the target image [14].

In this paper [2], the hub is on the examination of dynamic lossless data-embedding methods that allow one to embed [2] large amounts of data into digital images (or video) in such an approach that the unique image can be reconstructed [2, 10] from the watermarked picture.

## 3 Proposed Methodology

Mosaic image means that the image is produced by ordering many small blocks to form a vibrant larger image. The purpose of information hiding using mosaic images is a new way of securing information so that the data can be sent in a private and secure manner. The aim of this paper is to provide an efficient and better embedding algorithm as compared to all related previous techniques. The secret image is first divided into a number of blocks called "Tiles", and then these tiles are placed on the single image called "TARGET image" which we select from our database on some similarity measures with our secret image. We are creating our own database here to overcome the drawback of managing a large database. To create a mosaic image, we have to place each and every tile of the secret image over the target image with the efficient embedding algorithm on the particular region. The embedding of the tiles should be in the format. We are using a DCT algorithm for embedding.

It involves the following steps:

First, we take a secret image which is divided into tiles, and will convert each tile image into a binary value through 16 * 16 quantization, then we prepare a form of secret value by combining all the values, and then embed this secret value in the original image through 16 * 16 quantization, after embedding, we send it to the sender who will send the data to receiver, and at the other side, the receiver will extract the secret value using the inverse algorithm which can be seen in Fig. 2.

At the first stage of transformation, we must design and partition an image into the blocks. Each block is denoted by $a_i$, and then 16 * 16 quantization is applied to each one of the blocks denoted by $C_i$.

$$C_i = T\, a_i\, T'$$

The chip rate is calculated first, which is the ratio of total no. of pixel in the host image by the no. of pixel in tile, then the value of chip rate is used to find the $p_n$ sequence, which is an array initiated by 1 to the value of chip rate, and then temp is calculated by this chip rate and the value of temp rows and temp column is considered from all the pixel values of the cover image, the watermarked value of the pixel of the tile gets embedded in this region of temp rows and temp column.

At the second stage, quantization is the technique used in image processing to reduce the redundancies, and the lossy compression technique is used here which is achieved by compressing a range of values to an only quantum value. When the number of discrete symbols or the pixel are assigned by one constant value, and then discrete symbols which are generated as the pixels are less than this constant value

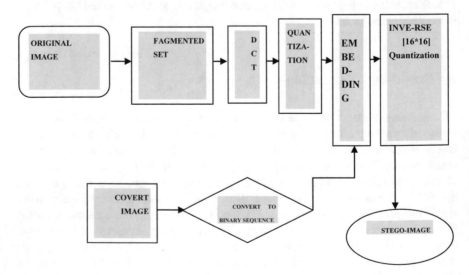

**Fig. 2** Outline of the proposed work

assigned to $-1$ and greater than this constant value are assigned as $+1$, and then by this, we are able to reduce the given stream, the stream becomes more compressed.

At the third stage, the secret image should be converted into a binary sequence ($m_i$). This binary sequence is embedded in the middle and high-frequency region of $C_i$ to get stego block $S_i$. Embedding process is based on applying magnitude modulation to the quantized value of the host $16 * 16$ quantization coefficient.

$$S_i(u, v) = \begin{cases} c_i'(u, v) + Qstep\ (u, v)/3, & \text{if } m_i = 0 \\ c_i'(u, v) - Qstep\ (u, v)/3, & \text{if } m_i = 1 \end{cases}$$

$m_i$ = each block of the secret image is converted into a binary sequence. Embedding process is based on magnitude modulation to the quantized value of the host $16 * 16$ quantization coefficients. Embedding requires the selection of a suitable region in which the tile should be embedded.

At the fourth stage, for each stego block $S_i$ (pixel value is converted in binary), the inverse $16 * 16$ quantization will be applied to get the output blocks $d_i$, and this is the inverse process of $16 * 16$ quantization as it requires to retrieve the original secret image.

$$d_i = T' S_i T$$

During extraction, the encrypted information needs to be extracted first, and then the correct sequence of tiles is obtained.

$$blockSizeR = 100;$$
$$blockSizeC = 100;$$

whole BlockRows = floor(rows/blockSizeR);
Now, each of the tiles generated is then converted into binary values using

$$dec2bin(message(i, j)$$

Here, we use pseudo-random number generation for the generation of a binary key from the image. The tiny image can be used for producing RNGs and also the image from the paint software. The pixel value of the image can be generated with the simple functions in MATLAB tool and it is converted into a string value. To convert the pixel value of the image into the binary value from integer, we have to check the RGB value of the each and every pixel, and then compare the pixel value. The corresponding values (0s and 1s) are written in the text file from left to right or in any other format. If there is a small change in the image, it leads to a big difference in the generated random numbers. Here, the concatenated value of the pixel is shown:

10101010101100100000000000
01000111110101110000000000
10100101111001000000000000
10010110010010010000000000
01101101111110100000000000
10010110111111010000000000
01000011111111010000000000
10111111110111110000000000
01010011010101111100010100
11101011111111111101110
10001011111111111111000010

From the generated binary value, embedding is performed, which embedded the secret image into the cover image and is sent to the receiver. At the receiver end, the binary value is extracted and is again converted into tile images. These tile images are then embedded to make a secret image (Fig. 3).

Here are some comparison measures, which are based on MSE, PSNR, NAE, NCC, and result analysis of the existing work and results analysis of the proposed work (Table 1).

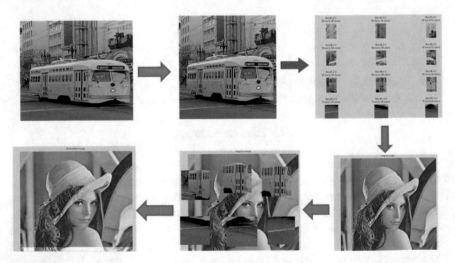

**Fig. 3** Illustrations of the creation of secret-fragment-visible mosaic image

| No. of tiles | MSE | NAE | NCC | PSNR |
|---|---|---|---|---|
| 6 | 0.22 | 0.0017 | 1.0008 | 54.79 |
| 12 | 0.21 | 0.0018 | 1.0011 | 54.78 |
| 15 | 0.31 | 0.0016 | 1.0012 | 55.79 |
| 18 | 0.29 | 0.0019 | 1.0010 | 54.3 |
| 20 | 0.30 | 0.0020 | 1.0004 | 55.9 |

**Table 1** Result analysis of the existing work and results analysis of the proposed work

## 3.1 Comparison Based on the Error Rate

See Fig. 4.

## 3.2 Comparison Based on the Computation Time

See Figs. 5 and 6.

## 4 Conclusion

The work that we proposed provides less compression time. This technique M-secure, here, provides a more efficient technique to authenticate the user to access the original image. By the different parameters, the work is more efficient with Ya-Lin Lee, and we have generated graphs and tables which is generated and tested through MATLAB.

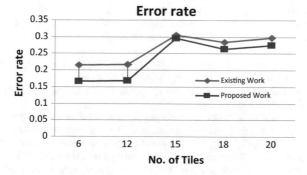

**Fig. 4** Comparative analysis of error rate

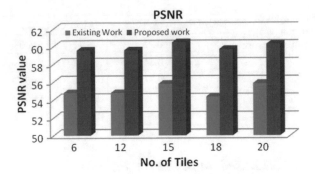

**Fig. 5** Comparative analysis of computation time

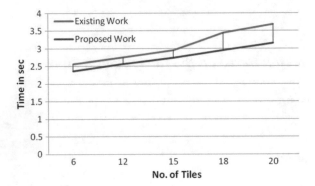

**Fig. 6** Computation time in s

# References

1. Ponti, M.: Image quantization as a dimensionality reduction procedure in color and texture feature extraction. Elsevier, pp. 1–12 (2015)
2. Rose, A. et al.: A secure verifiable scheme for secret image sharing. In: Second International Symposium on Computer Vision and the Internet. Elsevier, pp. 140–150 (2015)
3. Xu, D. et al.: Separable and error-free reversible data hiding in encrypted images, Elsevier, pp. 9–21 (2016)
4. Yang, C.-N. et al.: Extended color visual cryptography for black and white secret image. In: Theoretical Computer Science. Elsevier, pp. 143–161 (2016)
5. Qiao, T.: Steganalysis of jsteg algorithm using hypothesis testing theory. Springer Open J. EURASIP J. Inf. Sec. 1–16 (2015)
6. Bhandari, V., Tamrakar, S., Shukla, P.: A survey on: creation of mosaic images. In: International Conference on Advances of Electronics Computer & Mathematical Sciences (2016)
7. Chih-Wei, S., Chen, Y.-C., Hong, W.: Encrypted image-based reversible data hiding with public key cryptography from difference expansion. Published by Elsevier (2015)
8. Mishra, R. et al.: An edge based image steganography with compression and encryption. In: IEEE 2015, International Conference on Computer, Communication and Control
9. Zhou, Y. et al.: (n, k, p)-gray code for image systems. IEEE Trans. Cybern. **43**(2), 515–525 (2013)

10. Manjreka, A.: A novel approach for data transmission technique through secret fragment visible mosaic image. In: Emerging Research in Computing, Information, Communication and Applications, Springer (2015)
11. Edwina Alias, T., Mathew, D.: Steganographic technique using secure adaptive pixel pair matching for embedding multiple data types in images. IEEE, pp. 426–429 (2015)
12. Kuoa, W.-C. et al.: high capacity data hiding scheme based on multi-bit encoding function. Elsevier (2015)
13. Mishra, R. et al.: A review on steganography and cryptograph. In: 2015 International Conference on Advances in Computer Engineering and Applications. IEEE (2015)
14. Lee, Y.-L., et al.: A new secure image transmission technique via secret-fragment-visible mosaic images by nearly reversible color transformations. IEEE Trans. Circuits Syst. Video Technol. **24**(4), 695–704 (2014)
15. Li, X. et al.: A complete color normalization approach to histo-pathology images using color cues computed from saturation-weighted statistics. IEEE (2015)

# Analysis on Applicability and Feasibility of Dynamic Programming on the Basis of Time Complexity and Space Through Case Studies

Manish Dixit and Mohd Aijaj Khan

## 1 Introduction

In the world of competitive programming, dynamic programming serves as a powerful technique for solving mathematical problems to find the optimal solution [1]. To solve the problem, this technique works on decomposing the actual problem into subproblems and memorizing them. This technique is called memoization [2]. The values are stored in a specific location and when the program gets the input, it goes to that storage, access the value it has already computed, and outputs it to the user.

Memoization is the term, for example, in the real world to memorize the things. In the same aspect, the user analyzes the problem, and if he sees that the solution of the problem can be devised by dividing it into more subproblems and in some cases, further dividing, and then finally, computing it by using a suitable algorithm. In this process, the values have to be continuously computed to a limit and stored in some space. To get the desired output, the user has to access the suitable stored value and display it as the output.

Tabulation in dynamic programming is used to compute higher cases that build upon the lower ones till we reach our destination [2].

Dynamic programming works on the basis of these two pillars, tabulation and memorization.

M. Dixit
Department of CSE and IT, Madhav Institute of Technology and Science,
Gwalior 474005, India
e-mail: manishdixit@ieee.org

M. A. Khan (✉)
Department of Electronics Engineering, Madhav Institute of Technology and Science,
Gwalior 474005, India
e-mail: aijajkhan@ieee.org

© Springer Nature Singapore Pte Ltd. 2019
R. K. Shukla et al. (eds.), *Data, Engineering and Applications*,
https://doi.org/10.1007/978-981-13-6351-1_27

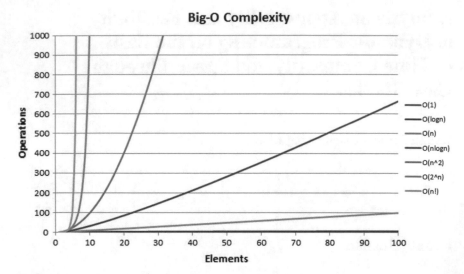

**Fig. 1** Graph plots the various time complexities based on Big O notation [4]

By analyzing this process, we see that the program has to run and store values consecutively and needs much space to store the values.

## 1.1 Time Complexity

Time complexity of an algorithm signifies the time required by it to complete its execution. A simple way of denoting is through the Big O notation [3].

Big O notation is calculated by counting the number of elementary operations in the specific algorithm. We always take the worst-case time complexity because of the difference in the performance of algorithms on different types of input data. So, the worst-case complexity is basically the maximum time in which the algorithm completes its execution on any type of input data.

Dynamic programming solution typically takes $O(n^2)$ or $O(n^3)$ for problems having an exponential time complexity (Fig. 1).

## 1.2 Time–Space Trade

The intuition behind dynamic programming is to trade time with space, so that we do not have to repeat calculations again and again if the logic says that the future outcomes are dependent upon the past ones [1].

So, instead of taking a lot of time to repeat calculations, we store the outputs and further build more outputs based on past outputs, store it as well to save our time later. The condition is that all the decomposed subproblems should be interdependent on past inputs.

### 1.3  Approach

In dynamic programming, to solve a problem, we have to solve all the subproblems which are related with each other, combine all the solutions of subproblems, and then reach a solution, which is essentially the global optimal solution [5].

On the contrary, in greedy algorithm, we compute the solution and select the choice which we consider optimal at the present time, not counting on all the solutions of further subproblems or forthcoming choices. Hence, we consider the local optimal solution as the global optimal solution [5].

## 2  Dynamic Programming Approaches

There are two approaches to compute and store the values in the dynamic programming technique, which are:

### 2.1  Bottom-Up Approach

This method in dynamic programming first computes the base case, and then builds further cases till we reach the solution of the mathematical problem [6].

For example, if we have an array, arx, then first, we compute the initial base case arx[0], and we start our approach to compute higher conditions till we reach our destination, i.e., arx[n]. We start with the lower cases and build up upon that to reach the optimal solution.

This is also known as tabulation. It finds the solution of every subproblem even if we do not need it.

For example, if the next stage is given as

$$arx[n] = arx[n - 1] * arx[n - 2]$$

## 2.2   Top-Down Approach

This method first calculates the highest state, which is the initial state in this case. We then compute the lower states till we reach the optimum solution [6].

For example, if we have an array, arx, we first compute the final case, arx[n], and we start our approach to compute lower states till we reach our desired destination, which is arx[0]. We start on the higher cases and build up upon to reach optimal solution.

This is also known as memoization.

An example of this is the calculation of factorial recursively and storing it in a data structure like array.

## 2.3   Comparison of Approaches

Tabulation is faster as compared to memoization because we directly have access to the past values from the data structure, while memoization is slow due to recursive calls.

All the subproblems have to be solved in tabulation even if they are not needed while we need not solve every subproblem and solve only the desired ones.

Critical thinking is needed to percept the conditions in tabulation and code get complicated if there are too many constraints. Whereas, it is relatively easier to memoize the solution as it requires recursive calling, but we have to be careful in complex problems as it may lead to overflow.

In terms of space, tabulation needs much more space than memoization as it needs to solve every subproblem to arrive at a desired case.

Every problem that can be solved by tabulation can be solved by memoization and vice versa [7].

## 3   Analyzing with a Case Study

Let us understand this with the help of a simple program of Fibonacci sequence, for easy understanding, we will use C ++ in Fig. 2.

Experiment result shown in Fig. 3.

Now here, we can see that the values are precomputed and stored and hence, accessed by the user.

```
#include<bits/stdc++.h>
using namespace std;
long long int arx[10001]; //defining the array

int main(void)
{
    int n;
    arx[0]=0;
    arx[1]=1;    //pre-defining values
    for(int i=2;i<10001;++i)          //pre-computing the values or memoization
    {
        arx[i]=arx[i-1]+arx[i-2];      //storing the values
    }
    cout<<"Which Fibonacci number do you want?"<<'\n';
    cin>>n;
    cout<<arx[n-1]<<'\n'; //accessing the value
}
```

**Fig. 2** Algorithm of Fibonacci series using dynamic programming

```
Which Fibonacci number do you want?
7
8

Process returned 0 (0x0)   execution time : 3.569 s
Press any key to continue.
```

**Fig. 3** Computing of seventh Fibonacci number using the above algorithm

On the contrary, if we solve it by recursion, we get Fig. 4,

Experimental result shown in Fig. 5.

It would still give the same output.

But the problem starts here that in recursive solution, the values are computed again and again. For example, when we know that fifth Fibonacci number is 3, why calculate again and again?

## 3.1 Analysis

In recursion, the problem is solved by:

```
#include<bits/stdc++.h>
using namespace std;
//long long int arx[10001];
int fibo(int x)
{
    if (x<= 1)
        return x;
    return fibo(x-1) + fibo(x-2);
}

int main(void)
{
   int a;
   cin>>a;
   cout<<fibo(a-1)<<endl;
}
```

**Fig. 4** Algorithm of Fibonacci series using recursion

```
6
5

Process returned 0 (0x0)   execution time : 1.743 s
Press any key to continue.
```

**Fig. 5** Experimental result

```
            fibo(5)
          /        \
      fibo(4)            fibo(3)
      /   \            /   \
   fibo(3)    fibo(2)       fibo(2)   fibo(1)
    /   \     /   \     /   \
 fibo(2)  fibo(1)  fibo(1) fibo(0) fibo(1) fibo(0)
/   \
fibo(1) fibo(0)
```

We can here see that fibo(0) and fibo(1) are calculated again and again [8].

Now, this is where dynamic programming technique comes to the rescue, and it trades space with its performance.

In terms of time complexity, dynamic programming solution gives a linear approach and a time complexity of O(n). By precomputing the series, we already have stored it in an array. Now the user inputs, and he gets the output immediately which is stored in the array, and now the complexity becomes O(1) which is a constant approach [8].

## 4    Case Study 2: Analyzing an Optimization Problem

This problem is featured in the Indian Zonal Computing Olympiad 2014 [9] (Fig. 6).

You are developing a smartphone app. You have a list of potential customers for your app. Each customer has a budget and will buy the app at your declared price if and only if the price is less than or equal to the customer's budget.

You want to fix a price so that the revenue you earn from the app is maximized. Find this maximum possible revenue.

For instance, suppose you have 4 potential customers and their budgets are 30, 20, 53 and 14. In this case, the maximum revenue you can get is 60.

---

### Input format

Line 1 : N, the total number of potential customers.

Lines 2 to N+1: Each line has the budget of a potential customer.

**Fig. 6**   ZCO 2014 problem 3

Constraints are: $1 \leq N \leq 5 \times 10^5$ (Figs. 7 and 8).

```cpp
#include<bits/stdc++.h>
using namespace std;
typedef long long int lli;
long long int arx[100001]; //define array

int main(void)
{
    lli n;
    scanf("%lld",&n);
    for(lli i=0;i<n;++i)
    {
        scanf("%lld",&arx[i]);   //fill in values
    }
    sort(arx,arx+n);   // O(n log n) complexity
    reverse(arx,arx+n);
    for(lli i=0;i<n;++i)
    {
        arx[i]=arx[i]*(i+1);   //use of dynamic programming
    }
    sort(arx,arx+n);
    printf("%lld\n",arx[n-1]);
}
```

**Fig. 7** Solution of this program by coding in dynamic programming (by myself)

```
8
40
30
20
50
100
42
160

Process returned 0 (0x0)   execution time : 195.051 s
Press any key to continue.
```

**Fig. 8** Experimental result

## 4.1   Analysis

As we can see from the algorithm, the program needs an optimum solution so as the seller can make the most profit by setting the price of the smartphone so that many of the users can buy it and he can make the most profit.

Decoding the algorithm in terms of the experiment,

*Input- n=6*
*People Bidding upon the smartphone: 40 30 20 50 100 42*
*Sorting it in O(n log n) complexity: 20 30 40 42 50 100*
*Reversing it: 100 50 42 40 30 20*
*By Dynamic Programming technique: arx[i]=arx[i]\*(i+1)*
*100\*1=100, 50\*2=100, 42\*3=126, 40\*4=160, 30\*5=150, 20\*6=120*
*Storing it in an array we have: 100, 100, 126, 160, 150, 120*
*Sorting it is same way in O(n log n) = 100, 100, 120, 126, 150, 160*
*So the maximum element in the array, we get: 160*
*Which is our answer*

In this way, by dynamic programming, we see that the optimum solution was found by first storing the values, or the subproblems and then using it once again in dynamic programming technique, i.e., $arx[i] = arx[i]*(i + 1)$, the array indexes are updated and we get a global optimum solution, which is the most desirable solution. Had it been greedy algorithm, it would have given correct solution for many cases, but not all, as it gives local optimum solution, and in competitive programming, where timing is most sought after, the solution would run to a Time Limit Exceeded verdict in large cases, i.e., large values of n.

## 5   Conditions of Dynamic Programming

For a problem to be solved by dynamic programming, it must observe the Principle of Optimality. The Principle of Optimality states that whatever the first (initial) stage is, the remaining decisions must be optimal with respect to the state following the initial decision. Dynamic programming computes efficiently by storing all the sub-results.

The original subproblem is decomposed to many subproblems. This division is done in such an order that the total result of the divided subproblems must be a polynomial or tends to be a polynomial. This is called Polynomial Break up, and is quite essential for a dynamic programming solution to be efficient [10].

# 6   Comparison with Other Algorithms

As stated before, dynamic programming finds the global optimal solution, whereas the greedy algorithm focuses more on local optimal solution. Greedy algorithm is more efficient than dynamic programming solutions. Dynamic programming is only efficient if there are not many partial results or subproblems to compute. Dynamic programming uses more space compared to greedy algorithm. Once the results have been precomputed, they can be accessed very swiftly just by going into the data structure's index, i.e., in O(1) time. Being more powerful than greedy algorithm, dynamic programming has a wider range of applications than the greedy algorithm.

In dynamic programming, the subproblems must have a relation with each other, means that first the initial result is calculated, and then the further decisions are related with respect to the initial decision. Then, the subproblems are combined to give an optimal solution. On the contrary, in divide and conquer algorithm, the subproblems computed are not related with each other. The result is combined, and the optimal solution is given [11].

As analyzed from the case study, dynamic programming used more space than recursion and only calculated a case once, stored it and developed further cases using the two penultimate and antepenultimate ones. Output was displayed by accessing the index in the array. So, by precomputing the values, we can get the result in a very short time. On the contrary, recursion calculates the subproblems again and again and uses more time but less space.

# 7   Expediency/Assets

It is no doubt, that dynamic programming is a powerful optimization technique that is versatile and has a very wide range of applications. A considerable amount of critical thinking is needed before we can classify a program as dynamic in nature. The concept of breaking down or decomposing a problem into many subproblems is a very desirable property of this technique. This idea enables us to reuse the stored computed results of subproblems as many times as needed, which is a very big advantage over recursion in which all the subproblems have to be calculated again and again. Dynamic programming is an essential tool to optimize a very complex problem by decomposing it to many subproblems, which are easier to solve and choosing the best-optimized solution.

The main strength of dynamic programming lays in the fact that the results of the computed subproblems are stored and can be stored for future references. The lowest answer usually finds out by a very efficient algorithm such as Quicksort, and it usually makes the run time lesser.

The part which makes dynamic programming attractive is that its rule of Polynomial Break up, which optimizes the performance, makes it more efficient by breaking down the problem and solving subparts.

Dynamic programming is such a powerful and promising tool that it is used tremendously in research in areas like Operations Research, Information theory, etc. [12].

## 8 Inexpediency/Limitations

Only optimization problems can be solved by using this technique. Many times, we need a local optimal solution, in which using a dynamic programming problem is undesirable. Complex problems are needed to be handled with some critical thinking to prevent overflow when recursive techniques are used, as in the case of memoization. It can be used on only a limited type of problems, like when the subproblems are not dependent on each other, dynamic programming cannot be used. The scenario of limited type of problems is due to dynamic programming's principle of following "Principle of Optimality" and "Polynomial Break up".

## 9 Applications and Research in Dynamic Programming

Dynamic programming is a mathematical concept and a powerful optimization technique that has a very wide range of real-world applications. It is very versatile and is used in many decision-making processes under unreliable risky conditions like optimum allocation of resources, risk management, etc.

Using this optimization technique, we can solve some particular aspects of domains which are stated below but not limited to:

1. Control Theory [13]
2. Information Theory
3. Competitive Programming
4. Operations Research
5. In Application Software, e.g., TeX typesetting system—calculation right amount of hyphenations and justifications [14].
6. Economics
7. Query Optimization and Decision-Making
8. Artificial Intelligence, Human–Computer Interaction, etc. [15].

In competitive programming and real-time applications, dynamic programming technique is used to optimize many problems like:

1. Longest Common Subsequence (LCS problem)
2. Longest Increasing Subsequence
3. Towers of Hanoi Problem

4. Knapsack Problem
5. Shortest Path Problem

   • Like Djisktra's, Floyd–Warshall and Bellman–Ford Algorithms

6. Manhattan Tourist Problem in Bioinformatics [16].


# 10    Conclusion and Future Scope

In this paper, we discussed how powerful dynamic programming technique is and how is it being used tremendously in research in various fields. Coming to conclusion we can now say that if we have to optimize a mathematical problem and find a global optimum solution, then dynamic programming is the best choice available in the optimization techniques.

Even though dynamic programming looks promising and attractive because of its functioning and advantages, this technique can still be improved by working upon its limitations, which is currently being done by computer scientists. More accurate techniques are being developed which will render dynamic programming a vintage.


# References

1. Larson, R.: A survey of dynamic programming computational procedures. IEEE Trans. Autom. Control **12**(6), 767–774 (1967)
2. Tabulation vs Memoization. http://www.geeksforgeeks.org/tabulation-vs-memoizatation. Last accessed 10 June 2017
3. Big O notation, Massachusetts Institute of Technology document, pp. 1−9. http://web.mit.edu/16.070/www/lecture/big_o.pdf
4. Big O Complexity Graph, http://down.admin5.com/demo/code_pop/18/95/. Last accessed 11 June 2017
5. Answer by Atiqur Rahman, Greedy Algorithm vs Dynamic Programming difference. Quora last accessed 11 June 2017
6. Dynamic Programming, Carnegie Mellon University, lecture 11, pp. 62, para. 3. https://www.cs.cmu.edu/~avrim/451f09/lectures/lect1001.pdf
7. Tabulation and Memoization. http://awjin.github.io/algos-js/dp/tab-memo.html. Last accessed 2017/06/10
8. Dynamic Programming I: Memoization, Fibonacci, Crazy Eights, Massachusetts Institute of Technology, lecture 18, fall 2009, pp. 1−6. http://courses.csail.mit.edu/6.006/fall09/lecture_notes/lecture18.pdf
9. Smart Phone, ZCO14003, ZCO Practice Contest, Codechef.com. Last accessed 11 June 2017
10. Bhowmik, B.: Int. J. Eng. Sci. Technol. **2**(9) (2010)
11. Dynamic Programming vs. Divide-&-conquer, Florida International University, Miami FL, pp. 1−11. http://users.cis.fiu.edu/~giri/teach/5407/F07/LecX6.pdf
12. Kapur, J.N.: On some applications of dynamic programming to information theory. Proc. Ind. Acad. Sci.-Sect. A **67**(1), 1–11 (1967)
13. Sahoo, Avimanyu, Jagannathan, Sarangapani: Stochastic optimal regulation of nonlinear networked control systems by using event-driven adaptive dynamic programming. IEEE Trans. Cybern. **47**(2), 425–438 (2017)

14. Answer by Slava Kim, What are some real life applications of dynamic programming? Quora.com last accessed 11 June 2017
15. Dynamic Programming, Artificial Intelligence. http://artint.info/index.html. Last accessed 10 June 2017
16. Huang, E.: Dynamic Programming and Greedy Algorithms, CS161 Lecture 13, Stanford University (2015)

Printed in the United States
By Bookmasters